A Trunk Full

Seventy Years with the Indian Elephant

NATURE, CULTURE, CONSERVATION

Series Editors

Mahesh Rangarajan
Visiting Professor, Department of History, Jadavpur University

K. Ullas Karanth
Conservation Scientist, Wildlife Conservation Society, New York
and Director, Centre for Wildlife Studies, Bangalore

Books in this series

Madhav Gadgil, *Ecological Journeys*
Vasant Saberwal and Mahesh Rangarajan (eds), *Battles over Nature*
Mahesh Rangarajan, *India's Wildlife History*
Valmik Thapar (ed.), *Saving Wild Tigers 1900–2000*
K. Ullas Karanth, *A View from the Machan*
Zai Whitaker, *Sálim Ali for Schools*
Gunnel Cederlöf and K. Sivaramakrishnan (eds), *Ecological Nationalisms*
Mark Baker, *The Kuhls of Kangra*
David Arnold, *The Tropics and the Travelling Gaze*
Dhriti K. Lahiri-Choudhury, *A Trunk Full of Tales*
EHA (E.H. Aitken) *Zoo in the Garden*

A Trunk Full of Tales
Seventy Years with the Indian Elephant

Dhriti K. Lahiri-Choudhury

permanent black

Published by

PERMANENT BLACK
D-28, Oxford Apartments, 11 I.P. Extension,
Delhi 110092

and

'Himalayana', Mall Road, Ranikhet Cantt,
Ranikhet 263645

Distributed by

ORIENT LONGMAN PRIVATE LTD
Bangalore Bhopal Bhubaneshwar Chandigarh Chennai
Ernakulam Guwahati Hyderabad Jaipur Kolkata
Lucknow Mumbai New Delhi Patna

ISBN 81-7824-166-8

Typeset in Minion by Eleven Arts, Delhi 110035
and printed by Pauls Press, Okhla, New Delhi 110020
Binding by Saku Binders

For Sheila, Toto, Alo and Sumitra

Contents

List of Plates and Maps

23. The captive is led along the Jhargram highway between Jatra
    Prasad (left) and Chandrachud (right).The captive seems to
    prefer Chandrachud's company.
24. The captive was sprayed with water every 10 kms during its
    35-km long journey to the kraal.

Maps

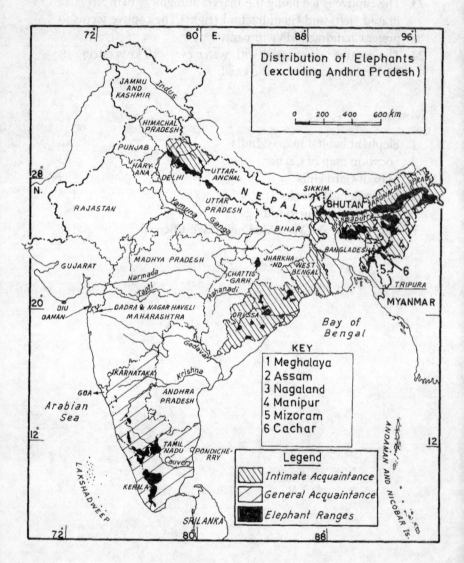

Elephant Map of India

Preface

I have had more than a seven-decade-long affair with elephants; I might even call it a magnificent obsession. This is not without precedence: my great-grandfather had it; my guru, Lalji, Raj Kumar Prakritish Chandra Barua of Gouripur, Assam, had it, and for that matter Akbar the Great, if name dropping may be excused, had it too.

The elephant has always been my passion but not my profession. I have done my share of surveys, writing of reports, have contributed my small share of papers—tinkering around with Excel program producing the usual graphs and standard-error calculations when presenting data—which few people will read, and fewer enjoy. Here I try instead to present some of my experiences. I had to be selective when writing them down, as too many images have been tending to clutter up my memory these past years. Looking back, I went through some interesting moments and tried to tackle problems which, at the time, appeared challenging. I hope I am able to convey to the readers the experience of some of these moments.

I perceive some distinct phases in my long acquaintance with elephants. First, coming to know elephants in general, both domesticated and wild; coming to know them as groups as well as individuals with distinct likes and dislikes, with temperaments and personal quirks of their own. For example, the big tusker Chandrachud, of the West Bengal forest department, would not hesitate to back into a closed lorry but another tusker, the great Jatraprasad of the department, would always go straight into a lorry, head first. Jatraprasad's temper was absolutely unflappable even under extreme stress. Once while carrying tourists in Jaldapara wildlife sanctuary, a mother rhino charged him frontally. The base of his curled-up trunk was split open, yet he stood as firm as a rock, as if aware of the responsibility he was carrying on his back. Numerous wild tuskers attacked him from time to time, sometimes within an inch of his life, but to his end he never flinched facing a wild tusker. Sharmila, a beautiful female elephant, is never jittery when facing a rhino or a wild tusker; but nothing would induce her to get into a closed lorry, which she seems to consider a deadly trap. Meghangini, another female bought by Subimal Roy of the forest department some years ago from Sonepur, had to be marched all the way from Sonepur in North Bihar to Jaldapara in North Bengal because of her antipathy to lorries. Nilkanta, an ugly-looking makna belonging to the West Bengal forest department, is a dangerous customer who has already taken the life of a few of his attendants and is prone to running amok with men on his back along with the mahout if anything startles him. Rajkanta, another departmental tusker, a fine-looking animal with thick, symmetrical tusks, just cannot tolerate motor vehicles. Once, chancing upon a jeep (with a senior wildlife officer of the Government of India sitting inside it) he smashed it to bits. The officer providentially escaped. Lal Bahadur, a big departmental tusker, as well as Nilkanta, was not approachable when in musth, but even a child, an apprentice pattawallah, could handle Jatraprasad with perfect ease when he was in that condition. The same was true of Lalji's Pratap Singh. Although all elephants are averse to touching dead flesh, there is a photograph of Pratap Singh in full musth in the Hollong forest rest house, Lalji riding him, women and children on his back, Pratap pulling out a dead leopard from the bush. The first section of this book, 'Elephants and Their ways', is about elephants such as these, elephants I have known as individuals.

The next phase concerns my experience of tackling rogues and marauders. Habitual killer-animals in the wild are rare, but they do occur. The worst are those that chase humans unprovoked and tear their victims limb from limb, even dragging the body around and scattering the remains over a large area. There is a special provision in the Wildlife (Preservation) Act for eliminating these animals. Some of these male animals can be tamed and trained with the help of modern technology. However, practically no state forest department in India except Karnataka has the necessary infrastructure of trained elephants and men to do so. These killer elephants tend to be large animals—I have seen such animals well over ten feet in height. The largest I have seen was 11'2" at the shoulder. Anything over 9' is considered beyond training, but Karnataka has successfully trained a number of elephants 9' and above, one actually 9'6".

Crop- and property-raiding adult male elephants, the major cause of man–elephant conflict, is another serious problem. Undivided Assam, which in the early days meant almost the whole of northeast India, and Coorg in South India, had special rules under the Elephant Preservation Act (1879) to 'control' the number of such animals. Similar rules were framed elsewhere in other British colonies as well, in Malaya for example. In Assam, eliminating proclaimed rogue elephants was permissible; elephants were also shot under the kheddah rules of Assam when trying to break down stockades built to capture elephant herds; in addition elephants were shot to control the number of marauding adult male elephants.

Elephant control licences, presumably following the model in Africa and elsewhere, were issued under rules framed under the Elephant Preservation Act. Not all animals eliminated under these rules were habitual crop and property destroyers. But marauding animals still remain the main cause of man–elephant conflict in India. Government inaction over this leads to indiscriminate retaliatory killings by humans. This is usually done by deliberate electrocution and poisoning, methods which are no respecters of sex and age. Thirty-seven such killings took place in Keonjhar district in Orissa alone between 1992 and 2002, a large number of the killed elephants being females and calves; and at least seven such killings occurred in West Bengal between 1994 and 2002, without agitating the animal rights activists. Scientists in India have recognized in recent years that keeping down the number of adult males is a necessary management tool; otherwise people tend to take the law in their own hands and apply it indiscriminately. 'Rogues and Marauders', the second section, is about delinquent animals I had to put down.

In the final section I talk about 'management' in the wild. I realized as early as 1975 that shooting *alone* was not a solution to man–elephant conflict—one must try to understand its root cause. This led me to study man–elephant conflict in North Bengal in 1975–6 where thirty to forty humans were being killed by elephants every year. Later this interest expanded to a study of the status and distribution of elephants in northeast India, covering Arunachal Pradesh, Meghalaya, Tripura, Manipur and Nagaland (IUCN/WWF Project 3031). From the mid-1980s to early 1990s, with the assistance of Ms Sujata Gurung, I undertook a study of man–elephant conflict in Meghalaya and on the south bank of the Brahmaputra in Assam; with Dr C.K. Sar in Orissa, I continued, adding a new dimension to the study I had originally started in North Bengal in 1975–6. Dr S.K. Acharya, Director General of the Geological Survey of India, carried out an elephant status survey in Manipur and Nagaland, an instance of exemplary meticulousness.

These detailed later studies of conflict were funded by the Liz Clairborne/Art Ortenberg Foundation and carried out with the help and support of the Orissa, Assam, and Meghalaya forest departments and the Ministry of Environment and Forests, Government of India. I am grateful for the unstinted support I have always received from the forest departments of West Bengal, Arunachal Pradesh, Assam, Meghalaya, Manipur, Nagaland and Tripura.

I am immensely grateful to the West Bengal forest department for accepting my participation in the various management operations carried out by them. These were extremely rewarding experiences, some of which I record in this book. I remember with affection and regard the late Dr V. Krishnamurthy whom I was privileged to have with me in some of these operations. I am grateful to Nitish Das for drawing the maps and Dr Shukla Bhadury for guidance and advice in the matter.

My editor at Permanent Black, Anuradha Roy, used a combination of coaxing and bullying that finally made me do this book. I can never thank her enough for her magnificent copy editing and chiselling that shaped my sometimes rambling narratives.

It is a momentous time for parents when their children start to talk good sense and provide advice worth taking, and my son Deep Kanta Lahiri Choudhury surprised me thus by going carefully through the text and making many valuable suggestions that tightened the narrative. Finally, this book would never have been written without the tolerance of my long-suffering wife accepting the hazards of living in forest bungalows, not always beautifully appointed or scenically well placed. She has made light of it all, wading through tall grass, clambering up to machans, and cooking meals with equal aplomb in the wild, always in an elegant sari.

Throughout two sets pronouns have been applied to elephants: the impersonal, neutral 'it' and the more personalized 'he' or 'she'. They reflect the quality of my experience: some elephants were to me just wild animals of the forest; some on closer acquaintance became 'persons' with distinct personalities of their own.

Everything I know about elephants I have learnt not in the classroom but by living with these extraordinary animals. I have tried recording my experiences with faithful adherence to detail and every attempt at accuracy. Omissions, if any, may be ascribed to the failing memory of an old man.

SECTION ONE

Elephants and their Ways

1

Growing up with Elephants

Children used to learn about the facts of life from birds and bees. Our wisdom in this area came from elephants. We eavesdropped as our elders, reclining against bolsters in the baithak khana, avidly discussed the private lives of our elephants.

Let us however begin, Alice-like, at the beginning and go through the middle, but no stopping at the end just now, if you please; for I cannot rest, at least not yet, all set as I am to sail beyond the sunset, to see in the Happy Isles the great Jatra Prasad, Chandrachud, Sher Bahadur, Shakti Prasad and Chanchal Piyari, all of whom I knew, and the equally great, if not greater, Banshi Bahadur, Nurjahan, Bholanath, Shambhu Prasad and Jung Bahadur whom I came to know of as a child, mostly as colourful chapters in family lore.

I might as well begin with my great-grandfather. At one time, the strength of his pilkhana exceeded thirty animals. Apart from duty on the estate, they were used in the dry months of March and April for shikar. (The recorded history of large shikar parties in the grasslands of swampy East Bengal goes back to the 1880s.) My great-grandfather earmarked the income from a substantial part of his landed estate for the maintenance of his elephants. Most of them were deputed to various rent-collecting outposts of the estate (dihis) as essential transport in areas without roads, which were liable to waterlogging during the rains. But all the elephants had to gather at the headquarters after the rains during the festival season of Durga Puja to take part in the ceremonial procession on Vijaya Dashami, the day our image of goddess Durga was immersed in the river. The more splendid the show that day, the greater the glory of the estate and its master.

Banshi Bahadur, the tusker, was my diminutive great-grandfather's favourite howdah elephant. The enormous height of his mount gave him an advantage over other sportsmen in the sea of grass they hunted in. Banshi was dead by the time I was a toddler, but his memory lingered on even when I had progressed to adolescence. I remember his huge hunting howdah which stood on the veranda of the malkhana, dwarfing half a dozen other such contraptions kept there in a row.

One of my grand-uncles who lived next door used to tell us how late one afternoon he heard the furious trumpeting of an elephant in the grounds of our house. Thinking something was seriously wrong, he rushed across to see what was going on.

This was during Durga Puja. All the estate elephants had come in from their distant outposts. That afternoon, freshly bathed, faces beautifully painted, they were lined up facing the high veranda where the Karta was feeding them dainties like rasagollas, bananas, and cut pieces of sugar cane.

Let us confess it was the festive season, and the Karta was not too steady on his pins, though it was not quite the time for the evening lights. He must have slipped off the high veranda holding out titbits to the pampered assembly. My grand-uncle found him on the ground under an upraised, sledgehammer-like front foot of Banshi. Banshi, standing on three legs with one foot raised, was screaming his great head off while the Karta was rolling about in the shadow of that foot, muttering slurred endearments to his beloved Banshi. Apparently, when the Karta had slipped to the ground before the startled Banshi, the elephant, in one reflex action, had pulled him forward to stamp on him; he was now having second thoughts, smelling a human. People were shouting: the Karta was going to be killed; but none dared to move to pull him away lest Banshi, surprised by a sudden movement, involuntarily brought his foot down. Eventually Karim Chopdar, one of the faithful retainers of the estate, shot forward and dragged him away, whereupon Banshi at last lowered his foot and went on trumpeting and hitting the ground with his trunk for quite some time. I might as well mention here that Banshi's score of human scalps already numbered six.

I also heard the story embellished with a lot of heroic gilt from Rahim Chopdar, Karim's son, who was among the spectators that memorable day. He never tired of telling me and the gathered assembly how as a young boy, he watched his father's magnificent heroism which saved the Karta's life that day from a dreaded killer elephant. This, he would say in passing, was why his family was gifted the acres of rent-free land they still owned.

Those days, in our part of the world, owners of landed estates lived in tightly packed clusters of houses, each with extensive grounds and representing a power centre. Each estate had its own people: senior staff, middle-rankers, underlings and hangers-on, and even its own prostitutes, who were housed slightly apart, as it would not have been correct for the staff of one estate to use facilities offered by a rival estate. They had an acute sense of belonging to their particular estates and were fiercely jealous of the privilege. And where privilege rules, can rivalries be far behind? Our sprawling string of power centres was no exception to this. In fact what was once a singly owned huge estate had been partitioned repeatedly over generations; all the resultant smaller estates were branches and taproots of the same old banyan tree, and the rivalry between them was therefore all the more intense, fuelled by the consciousness of imagined wrongs, and fancied slights, suffered in the misty old nawabi days, not to speak of more recent years.

Banshi Bahadur was made to be the key player in some of these parades of rivalry. The image-immersion procession on the day of Vijaya Dashami was one occasion when the pecking order among the Kartas of different estates tended to get sorted out. Tradition demanded that the images should be immersed in a flowing river, in this case, the Balua. The narrow passage to the river went through one part of the town where it was flanked by our Karta's property on both sides of the road. All Durga images had to pass through that passage on the day of immersion. Immediately arose the question of precedence. Whose image should go through first? Our Karta claimed that the Durga of his estate had the natural right to lead. The image belonging to the major rival estate therefore had to follow— it was so written. The latter tried to start their procession as late as possible to avoid a skirmish; but one cannot quite succeed when another party is actually looking for trouble. The late Hemantabala Devi, a daughter of the rival house, has written in her reminiscences how Banshi, with the Karta sitting on top, blocked the way describing intricate designs in the air with a long, brass-decorated bamboo pole in his trunk, displaying a skill which could be the envy of many a paik of the day. The tusker of the rival house, Banka Bahadur, mellow with age and blind in one eye, besides being nearly a foot shorter, was no match for Banshi, egged on personally by his master. The evening was Banshi's and there was no further dispute on that occasion about the right of way.

Intra-clan rivalries, frequently violent, were the norm; but when outsiders came in, a different picture was sought to be presented for the glory of the clan. For instance, on the occasion of a family wedding, when the groom's party came from a distant estate, rivalries were temporarily put under cover, and all united to impress and bedazzle the visiting party. All the available elephants of the clan were brought together to make up a spectacular procession to carry the groom and his party to their temporary lodgings for the happy occasion the following evening. All clan members and the Kartas joined in the procession.

On one such occasion, perhaps embittered by the unusually acrimonious Vijaya Dashami the previous year, our Karta decided to do things differently. All available elephants were duly sent for the procession, as custom demanded—including Banshi Bahadur. This caused a few raised eyebrows in the household. What would then be the Karta's mount? The Karta's inscrutable face, covered with a beard that would do a Professor Challenger or a W.G. Grace proud, discouraged further enquiry in that direction. Perhaps in his wisdom the huzoor was going to join the procession later in a palanquin.

Came the hour, and our Karta asked for one of his camels to be readied. Apparently he intended to join the procession on his camel, with its dazzling accoutrements, the animal a curiosity in the area and maintained because it was so exotic.

Now, our elephants were highly trained animals, the best of them responding to more than forty command words. They would not flinch from a charging tiger or a bull buffalo—but camels, strange creatures, were a different proposition. Timur Lang in his siege of Delhi had placed trussed-up camels in the protective ditch around his camp to deter the war elephants of the Delhi Sultanate, and won the day. I doubt if our Karta had Timur in mind but his idea was on sound Timur-track. The sight of the approaching camel created the desired effect in the long line of assembled elephants: total chaos. Elephants were running in all directions, many shaking off their mahouts, some smashing through the shops and buildings lining the main road, creating the mother of all mayhems, seeking safe haven from this strange-looking, strange-smelling apparition. At the end of the day there was the Karta very pleased with himself— nothing overtly hostile, all very correct, but achieving the desired effect nevertheless.

Bholanath, a massive makna belonging to my father's maternal grandfather, was almost certainly the tallest domesticated elephant in India in its days (an unbelievable 10'11"). He was as impervious to a charging tiger as to the attack of another elephant, even though the latter might be in musth. 'Be thou ever so big, I am bigger than thee' seemed to be his motto. He was like a great piece of rock which no storm could shake. Yet his end came as an anticlimax. Those days all owners of landed estates in the region lent their elephants for the great ceremonial procession in Dhaka on Janmashtami day. While crossing the river Buriganga, more a stream really than a river by East Bengal standards, Bholanath, at the head of the procession, sank without a trace into a bed of quicksand in full view of thousands of spectators. *Sic transit gloria mundi*, and so forth, but the legend of Bholanath lives on.

Shambhu Prasad, a great tusker (9'6") and a pilkhana mate of Bholanath, still remains for me the model of a really well-trained elephant. His temper in shikar has been casually recorded in a Bengali shikar book. A wounded leopard had sprung onto Shambhu's head. There was a shikari in the howdah. The leopard had bitten one of Shambhu's ears and was hanging from it, trying to get purchase with its hind legs to clamber up. As the narrator in the next howdah relates, Shambhu stood absolutely still, his huge ears spread out at right angle to the head, almost hiding the writhing body of the leopard, with only the top of the animal's head visible from the howdah. As the shikari in the howdah tried to aim at the leopard's head which was just visible over Shambhu's ear, the entire hunters' ring shouted urging him to be careful not to injure Shambhu—which shows that even in a crisis the priorities were not lost sight of. In all that excitement the warnings went unheeded, understandably perhaps, with only Shambhu's ear separating the leopard from the shikari. Meanwhile, the mahout, quite rightly, was hanging precariously by the dulshi on the other side of Shambhu's neck. A 12 bore rifle bullet went straight through Shambhu's ear, and that was the end of the leopard. The narrator describes how about a pitcher of blood spurted out of the big hole in Shambhu's ear (in retrospect, obviously a case of severed artery), and how eventually the mahouts heated the end of a gajbag to cauterize the wound and stop the bleeding. All this is narrated with the air of describing something routine. No untoward reaction, or even a hint of agitation is reported of Shambhu who, apparently, took the whole thing in his stride. Here, then, was the ultimate howdah elephant.

Nurjahan, from our own stable, according to contemporary accounts, was a thing of beauty, besides being a staunch howdah elephant. I have only a dim personal recollection of the animal, but remember how lovingly and vividly she used to be described by my seniors.

When my great-grandfather died, he left a rather capricious will. After an inevitable legal wrangle the estate was equally divided between my father and my grandfather. The first line of his last will and testament disposed of his only son, my grandfather, with a monthly allowance of thirty rupees and the right to free accommodation in the quarters of a second-ranking officer of the estate. The next line bequeathed his entire estate, totally and unconditionally, to my father who at the time was his only grandchild. Then followed three pages of detailed instructions on how his elephants, each mentioned by name, should be managed and maintained. As I have said earlier in a different context, it is all a matter of priorities. By the time the estate came to my father, three elephants had died. His share was seven elephants out of the surviving stable strength of fourteen.

Because of some family compulsions I spent my young days in the house of my father's maternal uncles. We engaged in some intense lobbying with the elders on Vijaya Dashami day to secure a ride in the ceremonial procession on an elephant of our choice. Unfortunately old Padma 'kani' (blind) inevitably fell to our lot as the most sedate and reliable animal available. Her faded glory, as a staunch howdah elephant from which many tigers were shot, did not impress us. It is obvious her's were not the qualities that readily appeal to young blood.

Even so, we would not miss the ride and at least four children under the guardianship of a rather large and smelly retainer festooned with water bottles and tiffin boxes, were packed into a charjama. I confess we were rather cramped for space, and the mile-long journey to the river and back was exciting but not very comfortable.

Elephants came into our nutritional regime as well. Lactating mothers with calves were brought into the andar mahal and were milked. Each one of us was made to drink about half a glass of the stuff, fresh from the teats of the animals. The thick liquid was universally hated by the children. It was supposed to be good for us, but, then, children seldom like what elders think is good for them.

When petrol was severely rationed in the early 1940s after the Japanese threat to north-eastern India, one of my father's uncles got round the problem with considerable ingenuity. He brought out an unserviceable, old open-body car of ancient vintage from the garage, cut off the front part, removed the engine and harnessed a young elephant to it. He used to move about town visiting relatives in that contraption. We were particularly fond of the young tusker Bhakta

Prasad, purchased from Sonepur a couple of years earlier, which had been put to the job. What a mad scramble there was among the children to get a ride on that elephant-drawn buggy.

Another of my father's uncles, much addicted to farming, in his case not a very financially rewarding enterprise, tried to get around the problem of fuel shortage by making a young elephant draw the tractor in his farm. It put the latest combines to shame as it not only tilled the ground but also produced and scattered manure in the process.

Shakti Prasad, a big tusker, is a character I remember very well from those days. He belonged to another of my father's uncles and used to be permanently stationed in the grounds of his house, next to which we lived. Shakti was of a rather irascible temper, not very tolerant of noisy, unruly children. One afternoon we were loudly admiring him from a road about 25 yards from his stall when he obviously misinterpreted our chattering as rude noises, and flung at us a large cut piece of a plantain stem meant to be a part of his ration. He narrowly missed his target but from then on we increased our observation distance by about 50 yards, keeping a wary eye on him all the time.

This uncle of my father was a great elephant fancier. He once purchased the entire catch of a government kheddah—numbering about fifty elephants—in the Rangamati area of Chittagong (now Bangladesh). This included a fine young tusker, which was named Sher Bahadur. Sher Bahadur, unlike Shakti Prasad, was even tempered, not given to throwing large pieces of plantain stems at noisy children. He fathered many calves in captivity, but never, reportedly, paid any attention to a cow not belonging to the original herd of captured animals to which he had belonged. This grand-uncle's solitary regret in life was that the strength of his pilkhana never reached the magic figure of one hundred. Some deaths in the pilkhana invariably intervened.

One of the annual great events I fondly remember was when some time before Durga Puja, a row of elephants occasionally numbering over fifty, belonging to my father's uncle, the owner of Sher Bahadur and Shakti Prasad, would be lined up before the simple, unostentatious, thatched bungalow of his father, my father's maternal grandfather. It was a glorious sight with calves frisking around, moving about under the bellies of the mother and their many aunties. Even now I owe to their memory a surge of sensations felt in the blood and felt in the heart.

Back home, when approaching adolescence, I remember for me the Puja festival really started with the sound of elephant bells approaching the house through the crowded bazaar beyond the large tank in front of the house, the inevitable pack of boys following. The elephant I really remember well, the remnant of the initial stable strength of seven, was Chanchal Pyari, a huge female (9'6") which had the reputation of being a staunch howdah elephant once. On the base of her trunk she bore the claw mark of a tiger that she had crushed to death on that occasion. My father was really a dog fancier, and did not care much for elephants. He kept only English mastiffs as a reasonable substitute, and did not add to the pilkhana he inherited. Besides, shikar parties were not his fancy either. He preferred the more exciting sport of Delhi politics, a sport more expensive as one did not make money from politics in those days.

Chanchal was later kept in the grounds of our house itself. Every day, returning from school, I flung my school bag away and rushed to her stall where she was receiving her evening grain ration in little packets made of plantain leaves. There is nothing more soothing to the nerves than seeing an elephant munching away contentedly on its little packets of paddy, salt and jaggery. I remember mother once caustically remarking that I'd never be anything but a mahout when I grew up. This, however, did not prevent her from having Chanchal brought into the andar mahal to feed her titbits with her own hands. I tried Chanchal for duck shooting in a beel (swamp) close to our house. She obviously had been well brought up as she never flinched when guns were fired from her back. This, however, once drew a sarcastic *sotto voce* remark from her old mahout: once tigers, and now whistlers, a sort of implied comment on the changing times and the decline and fall of things all over the place.

Though our own pilkhana had dwindled to a solitary animal, the pilkhana of my father's uncles still had about ten animals. They were all requisitioned by the government in the mid-1940s when the British wanted to set up an elephant brigade in Chittagong to counter the advancing Japanese. In fact, all privately owned elephants in the district were roped in for the purpose; not so in Assam, though, I believe. American-made, four-wheel-drive jeeps and monsters of transport vehicles and personnel carriers soon rendered the use of the elephant in the commissariat obsolete. Half of them died while in the Army's service, but Chanchal survived and the government acquired her for use by the forest department in North Bengal. Chanchal, I was later told, was the one animal which did not hesitate to cross the Torsa in spate. Eventually the contaminated flood waters of Jalpaiguri in 1967 brought about her end. It was a life grandly and gloriously lived.

We had no personal elephants after the Army took away Chanchal, and very soon no estate either to keep them.

But other gifts have followed, for such loss abundant recompense. I have since then known elephants in the wild, some of quirky characters, some truculent and vicious; I have been chased by irate cows, ignored by herds sometimes numbering fifty or more. I have made new friends in the forest department's pilkhanas. Madhumala meeting me on a lonely forest road at Malangi in Jaldapara, and gurgling or chortling with delight after recognizing me, was certainly adequate recompense for the loss of Chanchal Pyari. Jatra Prasad accepting me even when in musth, balanced out irritable, missile-throwing Shakti Prasad. I saw Lal Bahadur and Chandrachud grow up to become sturdy working animals, the latter turning out to be a fearless koonki, not afraid to take on any wild tusker. Of some of these animals, domesticated and wild, I sing below, albeit slightly off tune at places.

2

Lalji, the Elephant Baba

Rajkumar Prakitish Chandra Barua is unquestionably the greatest living authority on the capture and domestication of wild Indian elephants. It is as 'Lalji' that he is known to all who have anything to do with forests and wildlife, particularly elephants, in Northeast India. Second son of the late Raja P.C. Barua of Gouripur in Assam, he is the younger brother of Pramathesh Barua, famous film director and the leading *jeune premier* of the Bengali silver screen before the Second World War.

The Gouripur family was famous for their shikar expertise and their vast elephant stable. Lalji recalls a stable with forty-two elephants. He remembers one occasion when ninety-five elephants participated in the Governor's shoot. Elephants have been his lifelong love. As a boy of seven, he was riding home one day from a small-game shoot in his father's howdah, when his father asked him if he could ride an elephant. Lalji bashfully replied saying he couldn't. 'Would you like to?' 'Oh yes! Very very much indeed!' He was placed forthwith on the neck of Tara Rani, a sedate elderly female elephant known and loved for her good temper. He was then given his first lesson: prod behind the ears with your toes. The young boy was amazed to see how this huge animal responded to this subtle signal.

And thus, at the age of seven began a long, long affair which continues with unabated passion. Lalji is today sixty-seven years of age. His first elephant capture was in 1937 in the Buxa forest division in North Bengal. From then to 1981 he captured 1021 elephants in North Bengal alone. And, as he says, he is acquainted with approximately 10,000 elephants in captivity and probably as many in the wild.

Lalji is Lalji to the wide world but to true insiders of the elephant world, the traders, the phandis, the mahouts and their assistants, and the myriad camp followers, he is 'Baba'.

Lalji shot his first leopard at the age of nine and his first tiger at the age of eleven. In 1964 he shot his sixty-first tiger which was his last and his hundred and thirty first leopard. But all these were mere castaway leaves from a journal that could be called the 'yellow years'.

As a young man, Lalji's father sent him to the Presidency College in Calcutta for higher education. But he was unhappy in the city and wrote plaintive letters home pleading to be taken back to Gouripur for there were no elephants in Calcutta! Because of this lack of interest he did not get his graduate's degree. Neither he nor his father felt the loss or had any regrets. Instead of becoming one of the millions of graduates who overpopulate the Indian scene today, he became the one and only truly unique Lalji.

His regular routine from the age of twenty-two on has been to leave Gouripur after Kali Puja in October and move into the forest with his elephants till March 31. Then back to Gouripur with a month-long monsoon capture in June and out again in October. Since 1974, however, he has been living in the North Bengal forests practically all the year round with a two-year break between 1979 and 1980. Since 1981 Lalji has been assisting the North Bengal state forest department with his koonkis in the pioneering operation of chasing away herds of marauding wild elephants from destroying crops—a hazardous task which requires great skill and experience where two or three koonkis drive away a herd of over sixty animals.

One of Lalji's great loves has been Pratap Singh, a tusker which he had caught himself in a stockade. He recalls how it was love at first sight. Pratap was perfect in size, shape and every other detail. Lalji trained the animal himself with endless patience and love and discarded the coercive tactics that professional trainers employ to break the morale of a freshly captured animal. Pratap responded well and reciprocated the abundant love. The animal was fearless in shikar and rapidly learnt the meaning of 'bagh', tiger. Hearing the word bagh, Pratap would pause, extend his trunk to sniff the exact location of the tiger and would then move to the spot.

Once Pratap fell ill at a camp in the Bhutan foothills. Lalji was at Gouripur about fifty miles away and word was sent to him that Pratap was not well and not eating. This is a danger sign for an animal that normally feeds for twenty hours in a day. Distraught, Lalji took an oath at the feet of his family deity Mahamaya, and Ban Bibi, the guardian spirit of the forests, offering blood from his own chest if Pratap lived. He then crossed through the forest of Goalpara which in those days was teeming with wildlife of every description including elephants. The forest pathways were narrow dirt tracks. He drove all night breaking all obstacles to arrive at his elephant camp in the early hours of the morning. Pratap was well again and his appetite had returned. It was a tearful reunion. Lalji embraced his elephant and the wise, understanding, great animal, gently caressed his master with his trunk.

As dawn broke, Lalji entered the river nearby to keep his promise, his mannat, to Mahamaya and Ban Bibi. Standing waist deep in the icy water of the stream flowing down from Bhutan, facing the rising sun, he drew his hunting knife and inflicted a deep gash on his chest allowing the blood to flow freely and mingle with the water of the river. Then, cupping the water tinged with his blood, warmed with the rays of the rising sun, he made his offering to the guardian spirits he believed in.

Pratap was incredibly attached to Lalji and would fret, would not eat, when business called Lalji away. Lalji would leave his old, used clothes behind when departing for any length of time. There are unbelievable photographs that show Lalji riding astride Pratap's neck, the animal in full musth, with laughing children riding on his back; photographs showing Pratap, under Lalji's personal command, musth flowing freely from the animal's cheeks, extricating a dead leopard from a bush. Anyone familiar with the behaviour of elephants in musth will be able to appreciate this marvel. Pratap was allowed to mate with other domesticated cow-elephants that belonged to Lalji and he fathered five calves during his lifetime. Pratap, in full musth, was allowed to frolic freely and unfettered with two or three cow-elephants, his harem. Once a friend challenged Lalji to call for Pratap while the animal was engrossed in courtship. This was the supreme test of obedience. Nervous, lest Pratap let him down, Lalji called out to Pratap. Pratap responded at once to the call, left the cow-elephants and came to Lalji.

Finally, Pratap became a victim of the dreaded anthrax. Lalji stayed close to him. When he fell, his trunk rested against his master.

A few years later, Lalji went to Gaya and offered pinda for the peace of the departed souls of his father, his mother, and Pratap. A large and imposing oil painting of Pratap Singh hangs on a wall at Matiabagh Palace in Gouripur. Lalji donated a section of land to the government for a veterinary hospital in Gouripur in the name of Pratap. The government objected. 'An hospital in the name of an elephant!' 'Why not?' retorted Lalji, 'If there can be a college and a school in the name of my father and elder brother, why not an hospital in the name of Pratap?' The government retreated hastily in view of this logic. Pratap Singh Memorial Hospital stands today as a monument to a unique bond of love between man and animal.

Lalji has a clear, down-to-earth analytical mind. He is very familiar with Indian elephant lore, the so-called good and bad traits of an elephant, the unlucky signs, details of rituals related to the traditional capture of elephants, their training and also the traditional songs sung by the mahouts of Goalpara while training freshly captured elephants.

Jung Bahadur belonged to Lalji's father and was probably the tallest elephant in India through the twenties, thirties and forties. When acquired he was 10'4½" at the withers and eighty-plus when he died having been gored severely by another famous tusker of Gouripur, the ever-recalcitrant and cantankerous Shivji, who was notorious for his proclivity to charge other tuskers. Lalji really never cared for this huge tusker maybe because Jung Bahadur was well over a foot taller than Pratap; an unpardonable piece of impertinence for any elephant. I have often pitted Jung against Pratap, extolling the former's most outstanding points: his height and enormous thick, long tusks. Lalji dismissed his height. Pratap had a far better body structure, bandh. 'But what about the tusks? Jung's tusks were so massive,' said I, 'Pratap's were not even a quarter of their size!' A strained silence, an obvious sense of discomfort, but the retort would come fast enough: 'Large, yes, perhaps, but look at the shape! Pratap's were true palang, upturned, and symmetrical; Jung's were sloping towards the ground!' Almost a last-ditch fight for Pratap. Then a renewed offensive. After a pensive silence a sudden brightening up: 'What about their faces? Jung was *andharmukhia*, gloomy faced, but look at Pratap's face! So pleasant and smiling!'

Some of the famous elephants in Northeast India belonged to the Gouripur Raj. We have already met Tara Rani who was Lalji's first mount. She was pensioned off in her old age and released into the forest but would come back to her old stall very often. Then there were Jung Bahadur, Shivji, and Pratap Singh. There was Rudra Prasad, a huge makna, tuskless male elephant, standing 10'3" at the withers who never flinched before a charging tiger; Ful, a cow, picked up by Lalji's elder brother casually at a Calcutta court auction, a beauty in every respect; Uttam Prasad the killer of eleven people, and many more such stalwarts of the stable.

Kishen Lal, a tusker of noble proportions, ran away and returned four years later. This is probably the best-recorded instance of the proverbial elephant's memory. One morning in his stall at Gouripur, Kishen Lal suddenly decided to kill his mahout and realizing that discretion was the better part of valour, bolted for the reserve forests some forty miles away. Lalji gathered together half a dozen of his koonkis and went in search of the fugitive. For six months the vast sub-Himalayan forests of Goalpara were scoured thoroughly, hundreds of wild elephants encountered, but no sign of Kishen Lal.

Kishen Lal, a full-grown tusker, would not be tolerated in the vicinity of a herd by the established herd bull. It was therefore almost certain that Kishen Lal was leading the life of a loner in the bush. This entailed tracking solitary elephants—unpredictable and dangerous. There was a bizarre incident one day while following the tracks of a large bull on foot. Lalji with his group came to the banks of the river Ai and to his amazement and delight saw a big fellow wallowing in a pool of the river with his back to them. This could be Kishen Lal since a truly wild tusker was unlikely to be in the open river at midday. The bank of the river was steep here, and there was no immediate danger if the animal became aggressive. Supremely confident that Kishen Lal would not disregard his command, Lalji moved in close and standing on the high bank at the edge of the river shouted the command, 'piche', move back. The elephant immediately stood up and turned towards the intruders. Horror of horrors, it was not Kishen Lal at all but instead a huge, wild tusker in musth. Extending his trunk, he sniffed the air and charged till-tilt at the bank. Unable to climb up he gored the vertical side of the river bank viciously with his tusks. It then occurred to someone that if the elephant had managed to get to the river it could perhaps come up the same way as well. The search party then made it to a nearby village. A few minutes later, peeping from behind some huts, they saw that the elephant had indeed climbed up and was standing at the exact spot where they themselves had been a couple of minutes earlier, a picture of vicious aggression.

After six months of a futile search, Kishen Lal was given up as lost and the party returned to Gouripur. One morning, four years later, there was a general scare in the sleepy little town of Gouripur. A wild tusker had been seen coming along the Assam Trunk Road towards the town centre. He came straight to his former stall and went into a nearby marsh. He had the appearance of a wild tusker but the mahouts recognized him as Kishen Lal without much difficulty and he was encircled. One bold spirit leapt on to his neck and commanded, 'baith', sit. Kishen Lal obeyed at once and within a few minutes was tethered firmly without any resistance. He behaved as if he had never run away and he had not forgotten a single one of the forty-odd command words he had been taught. There were marks of healed scars on his body, some deep and long. Obviously, Kishen Lal had tried to fraternize with his wild brethren in the forest but had been rebuffed and was now back where he was sure of love and security.

Lalji's encounter with the famous Paglir Sahan, the mad woman's elephant herd, of Assam in 1954–5 is most authentic. He and his party were camping on the west bank of the river Saralbhanga in the Haltugaon forest division, now in the Manas tiger reserve. The entire day, on the west bank of the Saralbhanga, they had drawn a blank and it was now late afternoon. Next morning they were to move east across the river. Suddenly they heard sounds of elephants breaking twigs and branches and saw a Nepalese woman coming towards the river. At first they assumed she was a woman from a grazier's camp looking for her lost buffalo. But while throughout that day they had found no elephant, now, with her, elephants were everywhere. It was only then that they connected her with the elephants. The leader of the group was an experienced phandi, expert in noosing elephants. He asked the woman in Nepali who she was and where she lived. She merely smiled in reply.

Meanwhile the elephants had encircled the tree under which the party was camped. All were terrified except for the lady who was vastly amused with their reaction. There was no natural explanation for this. Was she then the 'Queen of Elephants' of the legends?

The men went down on their knees and prayed to her pleading that their presence should not be taken as a transgression of the laws of the jungle; that they were professional elephant catchers; and that they would seek her permission to catch elephants since their livelihood depended on it. The lady continued to smile without uttering a word. The pots of rice already on the fire had begun to boil. They offered freshly cooked rice to her on a leaf, lit incense sticks and waited with folded hands for her reaction. She put three or four grains of rice in her mouth and still smiling raised four of her fingers at them. Just then four of the seven koonkis accompanying the party appeared. This meant that the Devi had permitted them to catch four elephants.

Within a few minutes four wild elephants allowed themselves to be captured like cattle. They were still poised around the Devi, one of them a big tusker with asymmetrical tal-betal tusks: one higher than the other. The people bowed down to touch the ground with their heads convinced that she could be none other than Sahania Devi, the goddess of elephants. She then waved to the group and melted away into the darkening forests, followed by her elephants.

Soon the other three koonkis arrived. Lalji was against the capture of any more and refused to contradict the Devi's clear instructions. The others, temptation getting the better of them, rushed out on the three new koonkis to try and catch a few more. An hour later they returned to camp completely crestfallen. They had been unable to locate any elephant let alone catch one.

The last episode took place at Raimona, in the westernmost part of what is now the Manas tiger reserve buffer zone, in the Kachugaon forest division near the river Sankosh, the boundary between West Bengal and Assam. Three of Lalji's veteran phandis—Kalu, Mohan and Tirtha—took their koonkis to look for elephants. Only a few miles from the forest rest house they came across Devi, a 20- to 22-year-old Nepalese lady sitting under a tree. They recognized her, got off their mounts and offered her puja. Smiling, she lifted two fingers in what was by now to the veteran phandis a familiar gesture, a signal permitting them to capture two elephants. In less than half an hour they captured two and came back to where they had last seen the Devi. They performed their puja and thanked her. She acknowledged them and with a wave of her hand disappeared into the forest. They noted with wonder that though it had been raining her clothes were absolutely dry!

Every year at Sonepur in North Bihar starting on the full-moon day in the month of Kartick, a fair of animals takes place for a fortnight. This has been a tradition for a couple of thousand years at least. Cattle in tens of thousands, horses, ponies, buffaloes, goats, birds, camels and elephants, are all brought here and displayed for sale. Elephants, however, are the chief glory of Sonepur. Around 350 elephants are put up for sale every year. In 1978, the number was 550. All who have anything to do professionally or otherwise with the management of elephants must make it to Sonepur.

But visiting the Sonepur mela with Lalji is an unforgettable experience. The day is spent inspecting rows upon rows and more rows of tethered elephants, measuring their height, taking statistics of age, sex, tusker-to-makna ratio and above all talking to the various owners living in small tents opposite their animals. Baba is known to all.

When evening descends on the fair-ground and casual visitors withdraw, the elephant pattis—there are two of them—relax. The colourful drapes are off the backs of elephants and bonfires are lit using dried elephant-dung and waste fodder. A thick blanket of smoke and dust hangs over the area, the smoke keeping the mosquitoes away.

It is time to relax for the elephant men too. Lalji sits outside his small single-fly tent which is pitched with those of the other elephant traders. It is impossible to imagine his staying anywhere else when in Sonepur because he is as much a part of the Sonepur ambience as the elephants themselves. One by one the traders come in to pay their respects and seat themselves on empty gunny bags: Ramchunder of Barabanki, near Lucknow, Islam and Lal Babu of Bihar and other folk who for generations have lived with elephants. First, trade information is exchanged: the number of elephants captured in Dibang valley in Arunachal Pradesh this year; the rate of royalty in Tripura; the appalling low survival rate of calves captured by the Uttar Pradesh forest department; the possibility of capture permission in Karnataka, all this is vital to the trade. Being at Sonepur with Lalji one can gather information of the elephant world from the remotest corners of India. A buyer from Kerala dropped in. He was the secretary of the Kerala Elephant Owners Association. We were told the association was formed to combat the unreasonable demands of the mahouts who have a trade union of their own. An unusual conflict. He was at Sonepur to buy big tuskers, preferably over 10 feet, for the temples and religious processions. The height is important here because every inch over 10 feet fetches an extra three to four thousand rupees for a day's hire.

There is quiet excitement among the veterans when the bulk government buyers—the forest departments of the Andamans and Madhya Pradesh—are there. Prices run high, bulk purchases are expected. An old lady in saffron, an abbess, is camping under a tree nearby. She is here to buy a couple of elephants for her foundation and is presumably carrying on her frail person Rs 50,000 to Rs 60,000 in cash. Meanwhile, in front of Lalji's tent, the chillum moves from hand to hand, starting from Baba. Nerves ease and the mood becomes expansive. Inevitably, the conversation turns to the good old days. Memories and reminiscences are exchanged: which baiji was abducted by which Maharaja's mehfil or soiree, fond memories of Brij Kumar and Kumud Bahadur, the two most beautiful tuskers ever in the living memory of Sonepur, both from Tripura. Brij Kumar was purchased by the Maharajadhiraj of Darbhanga for Rs 25,000 when an average good tusker fetched only Rs 2500 or so. Deep into the night bonfires begin to die down and are stoked to keep them alive. Elephants are everywhere, some asleep on their side, some standing.

Lalji is one of the last of the disappearing generation. There has not been nor will there be another man like him, a man who has truly made elephants his life. Even in the days when he had a leisurely life with no stress or strain of earning a living or making a career, being a scion of a noble family, Lalji was unique. He, unlike his many sporting peers from the same social milieu, chose to make forests and elephants not a pastime but a passion and a way of life. Today, that way of life, the surroundings, the milieu, above all the forests with their teeming wildlife which made Lalji what he is, are gone. Only the memory lingers—glowing embers in a dying campfire, sparkling to life momentarily, fanned by the cool autumn breeze at Sonepur under a full-moon sky.

---

*Lalji died in 1988. This article was written in 1983 after the Sonepur mela of 1982.

3

Mammoth Love

Animals are not just bundles of instincts to those who come to know them intimately; each is an individual with personal likes, dislikes, and quirks. Conrad Lorenz has written with great understanding of animals and birds as characters. The bond between animals and their keepers or owners has been widely recorded. Kipling, for example, has described the bond between a tusker and its alcoholic mahout. No two elephants are alike, and one has only to know them well to come to this realization.

Take Vijay Singh for example, an ungainly elephant with short thin tusks, passionately attached to little Bala, the young female elephant in the pilkhana. The two were inseparable. One could not separate them without Vijay Singh throwing tantrums. Bala was everybody's pet, so sweet and tame that her mahout often neglected to tie her up at night. It was assumed that she would not stray far from Vijay Singh. But, alas, Bala was a bit fighty by nature, as sweet young things tend to be even among *homo sapiens*. One night a wild tusker came to the camp, and Bala, ready for a fling, ran away with her buccaneer suitor. Call it anthropomorphism if you like, but that is was what it was: a sudden irresistible pash for a short casual affair, a momentary urge to escape the boredom of the settled life at the pilkhana with a devoted partner. Vijay Singh, the big male, was chained in the stall nearby, and could do very little about it. He screamed, he roared, and tried his best to break his chain. Fortunately for Bala the chain held.

Next day all the pilkhana elephants set out in search of the missing Bala. Even after a week there was no trace of Bala in the vast sub-Himalayan forests around the pilkhana. The search parties were reluctantly called off. Bala was not only a sweet thing but also a valuable riding animal, fast, with a smooth gliding movement and a gently swaying pair of hips: no jerky movement.

Then somebody had a bright idea. Let Vijay Singh be set free in the forest. If anyone could bring Bala back, it would be Vijay Singh. On the third day, a battle-scarred but triumphant Vijay Singh emerged from the forest, preceded by a subdued and demure Bala. All was well once again. Bala seemed to enjoy her return to the staid security of the pilkhana and Vijay Singh's attentions. Only from now on there was no freedom at midnight for Bala.

Talking of male ardour among elephants, the story of Bajra Prasad, a magnificent tusker, was slightly different. Covered with a gorgeous caparison of zari and velvet, adorned with ornamental heavy anklets, long silver tassels hanging from his perfectly symmetrical pair of thick tusks, and carrying the state howdah of gilt and silver at the head of the Vijaya Dashami procession, he was the picture of regal dignity. Head held high when on the ceremonial march, he seemed to be fully conscious of his importance in the scheme of things.

But however regal he seemed, he always needed by his side the female elephant Raabi. They were inseparable whether in pilkhana, in processions, or engaged in more mundane tasks. Bajra used to be in musth for six months in a year, exuding maleness. Everything was perfect but for one little hitch: Bajra just couldn't do it. The problem was not one of virility. Bajra just did not know his way around the female body in spite of the mahout's guidance from on top; Nature had somehow slipped up in Bajra's case.

Jatra Prasad was different. He was the prize tusker of the West Bengal forest department, and his death is still lamented by all who knew him. In any tight corner, Jatra Prasad was the elephant of last resort. He was in musth about once in three years, and could be handled easily in that state. He never showed much interest in the number of departmental female elephants around and never sired a calf. In his youth he was once out grazing in the forest with Suryakanta, a sub-adult makna. Suryakanta was shortly afterward found in the forest with multiple injuries which could only have been inflicted by another elephant. A high-level departmental enquiry pointed the accusing finger at Jatra. It was held that Jatra had attempted to violate Suryakanta. A clear demonstration of the biological, rather than cultural, origins of homosexuality.

Malati, a sedate old female, was a placid soul, unflappable even in a crisis. Usually wild elephants rattle domesticated animals. Trumpeting, squealing, shivering, and even bolting are the normal reactions. Not true of Malati, though. She had been known to stand still, idly flapping her ears, when charged simultaneously by two rogue maknas. On another occasion a huge solitary tusker approached her for close inspection. She reacted by stretching her trunk to a succulent titbit high up in a tree, making her riders cling desperately for a few precarious moments to the ropes of the tilted riding pad. Cries of the riders in distress scared the tusker away. Malati, meanwhile, was supremely indifferent to the proceedings, including the attentions of a possible suitor.

Only sweetened coconut laddus, we found, evoked an emotional response in Malati. We had a small supply in camp which had gone rancid. We tentatively offered two of them to Malati and instantly accessed her heart. She was not greedy. She would take no for an answer only after two; for we had to ration our meagre supply strictly. To watch her munching these two tiny balls, savouring them rumbling with pleasure—a picture of perfect bliss—was almost to have had a princely repast ourselves. From then on, every evening when she was brought before us for her daily ration of grain, she would stretch out her trunk for the ritual two pieces of goodies; and there was a very happy and satisfied elephant before us.

One realizes late in life that the basic and more lasting instincts may go beyond sex—to rancid coconut laddus, for example.

4

Take a Makna by the Tail

One April morning in 1975, the Divisional Forest Officer (DFO), Jalpaiguri forest division (North Bengal) was about to leave for Darjeeling when a tea garden manager telephoned to tell him, with an ill-suppressed chuckle, that a wild elephant was making its way to Jalpaiguri town. The time being ten in the morning, and the place being Jalpaiguri, a flourishing district town miles from any forest, the DFO's first reaction was to attribute this 'news' to a warped sense of humour in the informant, a person whom the forest department loved not well, for reasons various and multifarious. However, to be on the safe side, he rang up his office to pass on the information.

Just as he was about to get into his jeep, the phone rang again. An extremely agitated and rather incoherent owner of a petrol pump on New Jalpaiguri Road, on the northern outskirts of the town, was trying to tell him that a wild makna had just walked through his service station towards the town, taking a good bit of the station's barbed wire fence with it.

This was serious, something one could no longer dismiss as a joke in bad taste thought up by a head still recovering from a riotous 'night before'. The DFO immediately informed the Deputy Commissioner (DC) and the Superintendent of Police (SP), and, cancelling his Darjeeling trip, started gathering up his own men.

Rushing to the spot where the animal was said to be, he found a sizeable crowd already assembled. The truant was standing next to a large bustee, demolishing at leisure clumps of banana. In a short while the DFO was joined by the DC and the SP, the latter with a platoon of armed policemen. Unsympathetic souls, as DCs and SPs tend to be on such occasions, they immediately asked the DFO almost in a chorus, what he proposed to do with 'his' elephant. As good bureaucrats, they were understandably anxious to settle, at the first available opportunity, the all-important issue of whose file the elephant was.

From then on, life was pure misery for the DFO. Every few minutes queries crackled in on the police wireless, from the commissioner and the Deputy Inspector General of Police; the crowd was swelling every second; and the elephant, a medium-size bull, was browsing placidly all the while on the banana plants, supremely indifferent to the furore all around.

It was, obviously, a critical situation; for, if the animal suddenly panicked and ran amok, with all that crowd milling around, it could mean a massacre.

Already a crowd more than a thousand strong had gathered there. Cycle-rickshaws were doing brisk business, carrying people from the town centre and back. Stalls had mushroomed, vigorously pushing dubious-looking sweetened coloured water, pan, cigarettes, and, of course, toasted groundnuts. A police cordon kept the crowd from pressing too close to the animal. It seemed to the harassed DFO that half the town was there, some with families, to witness this show of the year; and the nightmarish vision of the animal suddenly bolting through this holiday crowd floated spectre-like before him.

His gloom deepened as he brooded on the perfidy of men in general and tea garden managers in particular. Apparently, according to the latest intelligence, a maljuria group of three elephants had appeared that morning in the garden of the manager who was the DFO's first informant. The manager, then, in his wisdom, had organized a drive to push the elephants towards the town. It was partly pure fun and frolic, partly his way of drawing the forest department's attention to the problem of elephant depredation in his tea estate. Two of the animals broke through the line of beaters and went back to the forest. The third one, thoroughly confused, had found its way to the town, and was now standing, friendless and forlorn, Ruth-like among alien banana.

Around midday it was decided to drive the animal back to the forest, a good many miles away across the river Tista. A line of drummers and shouting men started beating the animal back towards New Jalpaiguri Road. For the time being the arrangement worked smoothly enough. The highway was reached without a hitch.

Just as the animal was about to cross the road, a lorry came tearing down from the direction of Assam. It had not occurred to anyone to block this arterial road with its heavy traffic of goods vehicles. It was then well past midday. The poor lorry driver, a bearded stalwart from the land of the five rivers, obviously in the throes of the traditional midday frenzy to grab lunch at the nearest dhaba, saw an elephant on the edge of the road, naturally took it for a domesticated one, and pressed the blaring horns.

As though on cue, the elephant wheeled round, scattering the beaters in all directions like chaff before wind. In no time at all, the situation was back to square one: the animal was in its favourite clump of banana near the bustee.

At this stage, the DC, a mild-mannered gentleman, by no means a man with a towering presence—his closest friends and admirers gave him a generous 5'2" in his socks—lost his nerve. Snatching a baton from a police constable, he started belabouring the bewildered lorry driver, a hefty fellow, standing well over six feet in his turban. For some time the atmosphere was chock-full of jigging batons, waving arms, shouts, and choice expletives in various North and North-east Indian dialects. Eventually calm returned, all passion temporarily spent on the hapless lorry driver.

It took some time to reassemble the beaters, and start drive anew. This time police pickets blocked the highway to give the right of way to the makna. The highway was crossed without incident; but soon there was trouble. The makna approached the railway lines running parallel to the road, a few hundred yards to its north, and immediately made up its mind that it did not like them at all. And that was that.

The beaters shouted like fiends, the drummers proceeded to beat the hell out of their drums; but the cussed animal remained impervious to it all. Even louder beat the drums, like maniacs screamed the beaters; the assembled citizenry with great enthusiasm contributed its quota of sarcastic remarks and contradictory advice; sedate housewives, who came on rickshaws to see the fun, delicately parted their veils with two fingers and said 'shoo'; but the makna, unmoved and unmoveable, continued to contemplate the railway tracks with disapproval, the ears going flap-flap, and the tail, pendulum-like, going swish-swish in a slow, unhurried rhythm.

The longer the elephant continued to be impassive, the more charged the atmosphere became. The DC looked distinctly like wanting to have another go at a passing lorry driver. The DFO, aware of the dirty looks that were being cast at him, had the uncomfortable feeling that he was being held personally responsible for all this. The situation had clearly reached a hiatus, and the much-harassed and profusely perspiring DFO was mentally preparing himself for a night vigil. Just then a youngster in drainpipes, which were then still very much the thing with the provincial sparks about marketplaces, came forward to ask if he could have a try. The young man mistook the DFO's bemused silence for a token of agreement. Aghast, the DFO saw the boy approach the elephant from behind, and before he could shout a horrified 'no', the thing was done. The fellow had grabbed the tail of the elephant and given it a vigorous tweak, the kind of straight, honest-to-God tweak a hard-working ploughman gives to the tail of his bullock when working his fields.

The makna shot across the railway tracks like a jack rabbit; the crowd, cheering, followed through the fields. Just before sundown it reached our old friend the manager's tea garden. As evening was approaching, the DFO decided to abandon the drive for the nonce, and leave the animal wherefrom it had started on its grand tour earlier on in the day. A purely operational decision, said the DFO, brushing aside with a benign smile the spluttering protests of the indignant manager, now a vivid purple in the face at the unexpected return of the prodigal.

Next morning, the makna, after his brief splash in the wide, wide world, was gone the way of all decent, God-fearing wild maknas: back to the forest and was heard of no more. I wish I could say here that the forest department and the tea garden manager lived happily together thereafter.

5

⌣

Gabbar Singh and Chomsky
in Dalma

They called him Gabbar Singh after the villain-hero of the film *Sholay*, the aggressive, firebrand tusker of Dalma, about thirty years old, beautifully built, with symmetrical tusks of medium length, always ready to take on anything that crossed his path, man or vehicle—a mastan if ever there was one. We knew him only as a loner, but no doubt he picked up his girlfriends now and then from the family groups moving around.

Mr S.P. Shahi, for long years the head of the forest department of Bihar, met him in Dalma in 1979. By then Mr Shahi had been well and truly bitten by the camera bug and had exchanged his rifle for a Nikon F2. He had also had a heart attack by then which he was wont to take as a minor impediment to his wildlife activities. Mr Shahi's introduction to Dalma came late, only after his retirement from service. I accompanied him on his first trip to Dalma around 1977, a cherished memory. There I was, a rank outsider, showing a former Chief Conservator around what was once his own domain. I was not with him however during his near-fatal encounter with Gabbar Singh.

I think it was some time around March 1979. He was being driven down to Bijlighati, one of his favourite water-holes, in a departmental jeep. Going past the first of the man-made reservoirs known as Barabandh, as he was driving down the twisting forest road, he saw a lone tusker in the water. The light was fine, and it was too good an opportunity to miss. He got off the jeep, and accompanied by the local forest range officer took his place on the raised rim of the reservoir, and started clicking away using a long lens. After a few shots, the slight metallic sounds of the camera's shutter disturbed the animal, and it slowly turned round to face the direction of the sound. It was then that the range officer recognized Gabbar Singh by the diagnostic broken tip of his right tusk, a scar of honour probably collected during a slight difference of opinion with a fellow mastan of the forest.

The ranger, a grass-roots forester, was well aware of Gabbar Singh's penchant for mischief, and in agitated whispers urged an immediate retreat. Shahi Sahib on the other hand had digested his Douglas-Hamilton well and was very hot on the idea of 'mock charge'. In fact, only a few months before while we were together in the Rajabhatkhowa forest rest house in Buxa, this was our recurring topic of animated dispute. My point was that the only way one could tell a mock charge from a real one was from the eventual outcome. A man killed? OK, it was for real. The man escaped: Ha! Ha! Obviously a mock charge. Little had I realized then that the theory would be put to test so soon.

To come back to Dalma, Mr Shahi ignored the advice of the range officer and went on clicking his camera—after all the light was so good and the setting near perfect. The tusker took a few steps forward whereupon Shahi Sahib calmly took off his tele-lens and changed back to normal—to get more of the background, as he genially explained to me later. Then, engrossed in composing his frames, he clean forgot that instead of a tele, he had a normal lens on. The elephant looked as large in the viewfinder as before. Plenty of distance, no need to worry, the jeep was only about 30 metres away. Then the tusker came forward in quick strides. Looking up from his camera, Shahi Sahib discovered that the assumed distance had suddenly decreased by three-quarter, and the fellow was only a matter of a few feet away. Pulled up to his feet by the range officer, he started running down the slope towards the jeep. He just looked back once and saw that the tusker had crested the rim of the dyke. Then he slipped and fell down, camera and all, and blacked out. The tusker came on. The range officer had the courage not to abandon his ex-chief. He stood by Shahi Sahib and started shouting and waving the white towel he was carrying on his shoulder. This made Gabbar Singh stop momentarily. He managed to pull up a groggy Shahi Sahib to his feet and both started running followed by the tusker. Shahi Sahib fell down once again near the jeep, but the sudden sound of the diesel engine of the jeep starting halted Gabbar Singh, and Shahi Sahib lived to comment on the miracle of his camera having escaped with only a slight damage, the only miracle he ever discerned in the whole business. Somehow he was distinctly less enthusiastic about 'mock charges' then on.

Next April saw us in a family group in Dalma. Dalma in late March–early April is all flowers and the bright green of new shal leaves. Kanchan of several colour, palash, the yellow silk-cotton tree (galgali as it is called locally) and shrubs of every kind all burst into flower together at this time of the year. The breeze is cool in the evenings. The lights of Jamshedpur from the TISCO bungalow on top of Dalma look like a carpet glowing with colour. The spirit of *joi de vivre* is everywhere.

On reaching Dalma we learnt that Gabbar Singh was still around, doing his very best to live up to his mastan image.

That was a year of drought. The man-made reservoirs on top of Dalma were much used by the fifty-odd elephants which annually retreat to the top of the Dalma Hills to spend their summer, very Raj-like. One of my objectives was to assess the status of the water-holes, some recently dug by the forest department to ease the stress of the drought on elephants. Things had advanced a lot since my first visit to Dalma around 1974 or so. The major water-holes now had concrete viewing boxes from where visitors could observe and photograph in safety elephants coming to the water. We chose the Majhela Bandh pillbox—actually a small square room of concrete—as our viewing point. Parking the jeep on the forest road, we climbed down a gentle slope to the box-like structure overlooking the water-hole. The snag was that the structure was suffocatingly hot on an April afternoon. Apparently this inconvenience had struck some earlier visitors as well; for we found a rickety ladder made of thin shal poles, struts tied with strips of bark, propped against its side. We, therefore, decided to perch ourselves atop the pillbox instead of venturing within.

It was nearing four in the afternoon when we were all comfortably settled on the roof—I with my cameras, my friend with a copy of Chomsky of all authors; and the others, including three children. I was convinced that it would be a futile vigil; for, with a copy of Chomsky around no self-respecting elephant could be expected to venture near us. Then we sat soaking in the ambience of the forest, the sound of dry leaves clattering down, the busy struttings of cattle egrets in the lush green grassy patches around the water-hole, the twitter of birds, the antics of a dragon fly on a dry twig overhead. Shutting his disciplined mind to all these inconsequential frivolities, my friend was immersed in his Chomsky, the picture of a raving intellectual lacking only the regulation straws in his hair.

Suddenly there was a flutter among the egrets, a branch broke noisily, and out of nowhere appeared Gabbar Singh on the edge of the water, his identity clearly marked by the right tusk broken at the tip. He plunged straight into the water and with all the enjoyment of a consummate voluptuary, started his bath. All elephants delight in water, and Gabbar Singh was no exception. Gabbar Singh was in musth as could be seen from the swollen temples and the tell-tale black mark of the oily discharge from the temples. We were happy with our cameras, safe on top of the viewing box, the jeep reassuringly close by. The children were excited, the ladies agog and even my friend condescended to put away his Chomsky for a while in honour of the visitor. The glorious afternoon clothed in new leaves and Gabbar Singh exuding musth and male vigour appeared to be made for each other. The frolic in the water went on with all manner of lusty squelches, gurgles, and swishes as well as whooshes. The elephant, besides being a noisy feeder, is also a noisy bather. He lay down in the water on one side, the tip of his trunk snorkel-like sticking out of the water, quivering slightly. Then he turned on his other side.

The tip of the trunk daintily rubbed the temporal orifice. He sat on his haunches in the shallow pool and wriggled his bottom about, rubbing it against submerged rocks—as thorough a cleaning as the finickiest of elephant mums could wish. He did everything in his bath except sing. Gabbar Singh, at peace with himself and the world, was totally indifferent to our presence, the clicks of our cameras, the excited ohs and ahs, the breathless whispers, the hissed admonitions, all barely fifty feet away and none of which could have escaped his extraordinarily sharp hearing.

The minutes rushed past, the sun's rim dipped, the shadows strode out. We heard the trumpeting of a calf some distance away. The local tracker accompanying us whispered that a herd was also approaching the water, and that it was time to leave, as soon there would be too many elephants about. As if to confirm the forecast, another shrill nasal trumpet rang out from a different direction. We had had our fill of thrills; it was definitely time to leave. I felt it was also time to show off my jungle wisdom and nonchalance. 'Nothing to it,' I said. 'We'll get to the jeep, but first let me shoo away the fellow,' and I clapped my hands wearing the casual, slightly superior smile of a man who has seen it all. Gabbar Singh completely ignored it; there was not even the slightest break in the steady rhythm of the flapping of his ears. I clapped again, and louder. No response once again. My superior smile was freezing into a silly grin. Another trumpet, this time closer by. Feeling this was no time for niceties of bush etiquette, I shouted at the damned animal, who acknowledged it with a momentarily cocked pair of ears and nothing more. It was getting dark rapidly.

With two ladies in elegantly flowing saris in the party along with three children of unbounded and irrepressible exuberance, and an insouciant Gabbar Singh barely fifty feet away, the jeep on the road up a rocky slope suddenly looked too far away. I shouted louder with the full power of a pair of lungs steeled by decades of lecturing. Gabbar Singh replied with a rude *brrr*, blown into the water with his submerged trunk. Then I invited all present to join the shouting party which they did with admirable alacrity and gusto. It appeared to have some effect on the closing herd—for the next trumpet in which one could almost detect a note of indignation, was from further away—but none at all on Gabbar Singh. Short of chucking Chomsky and the available Chomskyites at him, we had tried all the known tricks in the game on Gabbar Singh, but with no effect whatsoever. An unconventional tactic was obviously called for. As is common knowledge, elephants do not like being abused. I took resort to this next in sheer desperation. I confess, my style was somewhat cramped by the presence of the ladies and, especially, the children, and there was little help from my friend, Chomsky being of scant use in such basic matters. Whatever the reason, Gabbar Singh eventually decided that he had finished with his bath, and moved out of the water to a dusty pan on the other side of our concrete perch, increasing his distance from the jeep to a healthy hundred metres. Then he started dusting himself thoroughly, to the utter disgust of the ladies. Fancy dust and dirt, bucketfuls of them, after all that washing and cleaning! A bad example to the children!

That was all the concession that Gabbar Singh ever made to our protests. There was now a possibility of an organized retreat. The driver was asked to run to the jeep, start the engine, and keep it running. If the mastan came for the jeep, he was to drive away at once leaving us to a night under the stars. The tusker took no notice of the jeep's noise. Then the party in batches of two started making it to the jeep. While a group of two ran up the slope to the jeep, the rest kept a watch on Gabbar Singh, keeping up a loud running commentary on his behaviour for the benefit of those on the ground. Thus in slow and anxious stages we all retreated to the jeep and drove away, leaving Gabbar Singh to his toilet in the dustpan, with a vague feeling that we had been put in our place.

The next day was given to an inspection of the outlying water-holes. The range officer who had accompanied Mr Shahi the previous year kindly came to Dalma Top in his jeep to take me around. As there was no point in dragging the rest of the party along, it was decided that they would be left again at Majhela Bandh. We would return before dark and pick them up. Lightning never strikes the same place twice, I argued, and the same should apply to Gabbar Singh. As things turned out, I was right; what I had not noted, however, was that the adage did not apply to persons.

Depositing them on the familiar rooftop in the care of the

dragonflies, egrets, cicadas, and barbets, we proceeded in the range officer's jeep towards the water-holes on the northern face of Dalma. There is a ring road sort of jeepable track which takes one right round Dalma along the West Bengal face of the hill, cutting back through the range at one stage and emerging near Dimna on the other side, close to Jamshedpur. We were taking a 'short cut' from Majhela Bandh to this skirting road when suddenly one of the tribal trackers we had with us said, 'haathi'. We stopped at once. There was certainly an elephant in dense cover on the right side of the road, standing at right angles to the road. Try as we might, we just could not make out if the animal was facing the road or facing away from it, and, most important, if it was our old friend Gabbar Singh himself. One did not want to take unnecessary chances with Singhji. Ten minutes or more went like this; the animal standing absolutely still, no flapping of ears visible, always a bad sign that; and we no wiser. As this 'short cut' would save us quite a few miles and we did not have much time to complete our round, we hesitated to turn back; instead, we decided to have a close look on foot, keeping the jeep's engine running as a distraction for the animal. One tracker and I got off the jeep and tiptoed forward hugging the edge of the road and the fringing bushes. It must have been a very funny sight, but somehow no one actually laughed. I looked back once or twice at the jeep only to see a couple of grim faces behind the windscreen, and the completely blank face of the other tracker peeping out from the back. After advancing about twenty metres like this, the full outline of the animal emerged; and, low and behold, it was none other than G.S. himself, and facing the road to boot. The process of tiptoeing back to the jeep was considerably quicker; the last stretch, let me confess, becoming something of an undignified rush. There was now no choice. Our 'short cut' had been cut short by the big bully. It would have to be the longer route now.

Completing our inspection of the water-holes, we started retracing our way to Majhela Bandh. It was nearly five o'clock—enough time before dark to pick up our party. But Gabbar Singh meanwhile had had other ideas. Negotiating a sharp turn on that hilly forest road we suddenly saw him coming on straight at a clipping pace along the road with the purposeful air of one having an urgent appointment somewhere. He stopped seeing the jeep, and we gladly reciprocated the gesture. I could not help observing bitterly that it seemed one could not move a foot in this blighted Dalma without tripping over this bully. I was developing a definite feeling of being persecuted.

However, a little thing like a jeepload of people was not going to hold up for long a mastan like Gabbar Singh. Sure enough, after a few seconds he was on the move again, coming straight towards us. The right of way he obviously considered was his—come to think of it, strictly speaking, so it was too. We promptly started reversing— there was simply no space to turn the vehicle around on that narrow, hilly, forest road. This went on by fits and starts for nearly half a kilometre—a nerve-racking business on that kind of road. Every time the engine of the diesel jeep was revved up, the tusker halted; as soon as the engine fell back to normal, it started coming again. Reversing all the way, at last we reached a wider strip and could turn the jeep around with some effort, the tusker closing in every second. With the jeep turned round, we were now in a position to outrace it—one relief. After proceeding down the road for another 200 metres or so, the tusker following us steadily all the while, I asked the driver to stop the vehicle, keeping the engine running of course, so that I could climb out of the front seat and take a steady snap from the ground of our relentless pursuer. As soon as I was out of the vehicle and standing on the road, exposing my separate identity, the tusker stopped in its tracks and fanned out its ears—a sight and a sign that never fail to quicken the pulse of an elephant fancier. It was about 50 metres away. I focused my camera and clicked the shutter. The instantaneous reaction to the metallic click was a headlong charge. The massive head slightly cocked to one side and kept low, the trunk curled in, one piggy eye keeping me fixed in sight, Gabbar Singh came for me like a madly speeding killer Calcutta minibus. As if on a spring I turned towards the jeep, but by then the driver had started to speed away. I somehow managed to fling myself into the jeep. Within seconds at least 500 metres separated us from that great big hulk of mischief. After our previous evening's experience, it was as if Gabbar Singh was teaching us the tyros, a lesson in how to shoo away undesirable elements. Craning our necks back we could see the fellow no more. We stopped again. Time now for a decision. If we wanted to avoid that road—Gabbar Singh now disputing our right of way—we would have to circle down all the way to Dimna, then take the Jamshedpur–Ranchi highway, and come up the hill from the other side. If we did that, it would be at least eight or so in the evening before we could hope to reach our party on the water-hole. While they would be perfectly safe on the concrete platform where we had left them, they were unlikely to savour the hours of waiting in total darkness without any food or even a torch, and not knowing what had happened to us. Discretion suggested the longer Jamshedpur–Ranchi route; but there was also the nagging suspicion that one would never hear the last of it and that the story of how old D.K. abandoned his friends and family in the forest of the night in vain pursuit of idle pleasures would be dug up from time to time to make irrelevant points at inconvenient moments. One has to keep one's cool in such crises, and I was determined to keep mine. It was a matter of choice, the very stuff of Greek Tragedy. Firm in my resolve to avoid any error of judgement, I decided to give the old route another try. But first we had to find out what Singhji was up to just then. We again turned the vehicle around, or to be precise, persuaded a very reluctant driver to do so, and proceeded towards the spot where we had been seen off by Gabbar Singh. There was no sign of him, which did not mean anything. Was he again up to his usual prank of waiting for us in dense cover by the wayside in gleeful anticipation? Knowing by now some of his ways, it was not only possible but also very probable. The only thing was to find out the ground situation on foot. The tiptoeing act was put on again—only this time I chose to opt out of the cast, pleading a bruised knee sustained in the course of jumping into the jeep.

Everybody was sympathetic, as they could see that I had not been built by Nature for such stunts as taking flying leaps into fast-moving jeeps. Our two trackers silently went forward for about a hundred metres, one of them with his head down looking for spoor on the dusty road, we tensely watching their every movement. They abruptly pulled up at a spot, crouched down, moved to the side of the road where the ground sloped down towards West Bengal, watched the side of the road intently for a while, and then came back to the jeep in hurried but silent steps. Yes, Gabbar Singh was there all right, but not close to the road. He was standing under a mango tree some fifty metres down the slope. This was good enough to race past him even if he decided to come for the jeep once again. The plan worked. We were back in Majhela Bandh in time.

I never saw Gabbar Singh again. A few years later when I was back in Dalma, they said Gabbar Singh had been shot in West Bengal as a rogue; but I could never actually relate him to any particular rogue elephant shot in southwest Bengal. Maybe Gabbar Singhs do not die, they just fade away.

Why was Gabbar Singh so aggressive? Apart from his musth in the present case, the answer I think lay in his half-broken right tusk exposing the nerve cavity. The entire inside of the remaining part of the tusk, the bonesocket of which extends into the skull almost up to the eyes, must have become infected, and the animal must have been in agony all the time. Just think of what a bad toothache does to one's temper, and judging by the size of his infected tooth Gabbar Singh's must have been the 'mother of all toothaches'. I have known such animals to turn rogues. Once such an animal had to be destroyed. During the inspection of the carcass it was discovered that the entire tuskcavity was chock-full of maggots. And, looking back, if we resented bumping into him every time we moved out in Dalma, he on his part had all the reason to object to our popping up everywhere in his path. At least we did not have the toothache— Gabbar Singh did. Poor, poor, Gabbar Singh!

6

The Saga of Harjit

Harjit was a gigantic solitary makna in the prime of life, about 10'3" at the withers, and reportedly blind in the left eye. The Bengdubi Army Camp had christened him 'Harjit' after one of the past camp commanders who reportedly had the same physical handicap.

His range was the Kurseong division forests of North Bengal, west of the Siliguri–Darjeeling road. Here he moved like a liege lord with the air and assurance of one owning the place. For years this one-eyed buccaneer dominated the life of the local people, extracting grudging seasonal tributes. Over the years he had become a kind of an emblem of the place, and was unquestionably the most talked-about personality of the region.

For a long time the army supply depot at Bengdubi was his favourite foraging ground, till elaborate elephant-proof barriers denied him his prerogative. A visibly shaken army officer once reported to me that the elephant was white, a 'safed haathi', in fact. 'I tell you, Sir, saw the damned thing myself in the morning. Milk-white, I tell you, and it wasn't Sunday morning either.' Apparently, the previous night Harjit had broken into a warehouse storing flour and sugar. The officer saw him in the morning emerging from the place, covered with the stuff, and was never the same man again. When white elephants begin appearing, can pink ones be far behind?

Harjit was certainly not the shy and retiring type. He seemed to seek the forests more for shade than cover. His total disregard of human presence occasionally led him to escapades wholly regrettable.

Lalji was once camping at the forest rest hut at Bamonpokhri in Kurseong division. One day at the incredible hour of two in the afternoon, the people in the hut suddenly saw an elephant's trunk snaking in through the open window. Hastily all backed against the opposite wall, as the huge proboscis went unerringly for the bunch of bananas hanging in one corner of the room. Moments later the indignant company saw their next morning's breakfast disappearing through the window.

A few months later Harjit took to visiting Lalji's camp daily. There he was one day, at about four in the afternoon, determinedly approaching the hut. Everybody shouted, everything metallic—pots, pans, buckets, and all—was beaten lustily, without any apparent effect. Finally, Lalji himself, armed with a borrowed single-barrelled shotgun loaded with no. 6 shots, took his desperate last-ditch stand on the veranda, barely thirty feet from Harjit, in defence of his bananas. It was a battle of nerves. They waited face to face in tense silence, each waiting for the other to make the first move. Not a single flap of the huge ears all spread out, not a twitch of the tail. After a long wait Harjit changed his course, pretending he had intended to go the other way all the time.

Once in the dead of night there was a frantic telephone call from the house of the vice-chancellor of North Bengal University to the local police station. Harjit was on 'dharna' in front of the portico. It took hours to dislodge him from there. Tough is the lot of vice-chancellors these days all over the country, but never was it tougher.

This sort of thing, obviously, did not help to improve Harjit's public image.

Understandably, Harjit was the *bete noire* of the elephant squad of the forest department, recently constituted to combat elephant depredation. Their effectiveness against Harjit raiding crop was nil. Spotlights focused on him were ignored; flaming torches thrown at him were either stamped out or thrown back. Shotgun pellets fired at his legs caused only a muscular twitch, the sort of twitch one sees on a cow trying to dislodge a fly on its back.

One winter the squad's credibility received its severest drubbing from Harjit. It was a freezing Dooars night, and the old sinner, as was his wont, was on crop. After battling valiantly, albeit unsuccessfully, for hours to drive him off the paddy fields, the squad retired in frustration to a nearby village and lit a fire to warm themselves, attended by a jeering crowd markedly deficient in the milk of human kindness and that heavenly gift, sympathy. In a few minutes, Harjit appeared on the scene and appropriated the fire, whereupon all scattered like chaff before wind. For the next couple of hours or so, Harjit stood there soaking in the warmth, while the shivering squad members watched helplessly from a distance. The cup of their ignominy was full.

Harjit's ruling passion in life was food and, let us face it, for all his charm, Harjit was a dedicated gormandizer. Once the urge for female company came upon him, and he decided to abduct a female elephant belonging to a herd which was passing through. This action of Harjit, who appeared as brazen about love as about food, was resented by a bull temporarily attached to the herd, an enormous tusker. A night-long battle of the giants followed. In the morning, a terribly battle-scarred Harjit was forced to withdraw from the arena. For several weeks there was no news of him, and people began to wonder if Harjit had at last paid the wages of his accumulated sins. But one evening he was back, on crop, as usual.

Harjit might have been a glutton, but he bore no malice towards man. He was no killer. Actually, there was no established charge of deliberate and unprovoked manslaughter against him. A story goes that in some village near Panighata on the border with Nepal, he knocked down a hut to get at the stored paddy within. The inmates of the hut fled leaving an invalid old man behind. The elephant stood within a few feet of the groaning man demolishing the paddy, but not touching him. On another occasion, a near-blind old man was trying to protect his crop by shouting and waving a flaming torch about. After a while Harjit, for it was none other than Harjit who was his guest, finding this bobbery distasteful, resolutely approached the man, took the brand away from his hand, threw it away, and calmly went back to his supper. It was minutes before the old man realized what had happened.

Harjit, then, was no hater of man. His was a case of not loving man less, but loving paddy, maize, flour, banana, salt, and, of course, country liquor, a hell of a lot more. When pestered by people he made impressive demonstrations of threat, purely, it seems, as a matter of form; for, as a rule, these were never pressed home.

I saw him last in June 1977 close to the Pankhabari road near Rakti Jhora bridge, a massive bulk standing nonchalantly among the tea bushes. It was about nine in the morning, by which time elephants with any respect for elephantine custom and usage retire to the depths of the forest, far away from the madding crowds of smelly bipeds. We got off our vehicle, approached him close, and got ready the cameras, taking a wary note of the ominous wet 'musth' patch on the temple.

After the first click Harjit turned to face the camera. Soon, however, he felt reassured, and went back to a contemplation of whatever it is that elephants contemplate standing among tea bushes at such hours of the day. And so it was past an hour before long. Meanwhile busloads of people, cars, army vehicles carrying jawans, lorries and tractors overflowing with tea garden workers, all intent on enjoying the fun, had piled up on the road. The roadside was fast assuming the character of a village fair. Harjit, despite his habitual insouciance, was getting restless.

We decided to break away from the crowd, and went to Bamonpokhri to fetch Lalji. When we returned at about eleven, Harjit was still there. However, the sun was getting warmer, and Harjit decided to move on to thicker shade across the river Rakti.

It was an unforgettable sight, the great bull stalking majestically across the bare dry bed of the river, his towering height, now clear of tea bushes, fully exposed. A quick 'whoosn' in midstream, a momentary pause on the fringe of the forest across the jhora, and the vision in reddish brown was gone, leaving a strong feeling of 'bird thou never wert' in us.

As has been noted before, Harjit's besetting sin was gluttony or 'gula', one of the seven deadly sins, with an occasional dash of lechery or 'luxuria', another of the deadly ones. Further, he had allowed his 'gula' to cloud his reason, to corrupt his will. He, alas, thought nothing of knocking down huts to get at the stored foodstuff within. In 1976, he reportedly damaged over forty huts in one night. In July 1977, he again damaged a few. Wages of sin being what they are, the death sentence was passed on him. Harjit had gormandized once too often. When Harjit had to go, it was a melancholic day. With him went a touch of drama which coloured the everyday life of the area. The Kurseong division forests have never been the same again. It was a true tragedy of character, the tragedy of one who ate not wisely but too well.

SECTION TWO

Rogues and Marauders

7

Rogues and Marauders

Killing elephants was never considered a sport in India, except by some tribal communities in south and northeast India; but catching elephants was. According to one interpretation of Arrian's account of the Macedonian adventurer Alexander in India, the invader also practised this sport for relaxation between campaigns in northwest India. It was so considered by Akbar's court as narrated by Abu'l Fazl. Tipoo Sultan of Mysore also attempted it, but it is not clear if this was for his army's needs or for sport—perhaps for both. (Sanderson 1878). In the nineteenth century this sport was practised in the princely states of Korea and Sarguja in present-day Madhya Pradesh. This royal sport was also practised in Nepal, where elephant shooting was strictly taboo on religious grounds (Smythies 1942), and also further east in nineteenth-century Siam (Whitney 1905). Elephant kheda in Mysore during the Raj and even after was also a grand spectator sport, witnessed by members of the English royalty and viceregal guests.

In common parlance in northeast India one talks of 'bagh' [tiger or leopard] shikar' [blood sport of killing an animal], 'bhaluk [bear] shikar', 'pakhi [bird] shikar', 'peti 'or 'mela' shikar [capturing elephant by noosing]', 'kheda shikar' [elephant capture by driving them into a stockade]', but when it comes to hunting an elephant it is just 'mara' or killing, not shikar or sport. The Maharaja of Cooch Behar's shikar party accounted for 365 tigers, 311 leopards, 207 rhino, 438 wild buffalo, and countless deer in the 36 years between 1871 and 1907 in Lower Assam and the Dooars, but not a single elephant in an area teeming with wild elephants as it is even now (The Maharaja of Cooch Behar 1908). The party of Maharaja Yudha Samser Jung Bahadur of Nepal shot in eight seasons from 1933 to the early months of 1940, 437 tigers, 73 leopards, 53 rhinoceros, but not a single wild elephant. When two tuskers started raiding the shikar camp at night, they were captured, a process which provided rare entertainment for the royal party. On royal command a young mahout on a mount from the royal stables took on one of the giant marauders in full view of the party gathered on the edge of an escarpment to watch the show below. The young man lost his life in this unequal combat— but there was no religious taboo against the killing of mahouts (Smythies 1942). Between 1896 and 1970 the shikar party of the Raja of Gouripur, Assam, accounted for over 500 tigers, but not a single wild elephant in the heart of northeast India's elephant country (Game Books of Gouripore, unpublished). When it became imperative to destroy a dangerous animal, the job was given to a trusted retainer. As one of the sporting sahibs from North India once observed, 'elephants were something one shot from, not shoot at'.

The sahibs were the first to start shooting elephants as sport some time between 1807 and 1825, probably in Sri Lanka (then Ceylon). Bounty hunting of elephants irrespective of age and sex soon started in South India as well. The official policy of giving a financial reward for the destruction of elephants encouraged the shooting of thousands of elephants in Sri Lanka. The rate announced by the government in India was Rs 71 per tail. This imperial tradition of encouraging bounty hunting of 'vermin', with the tail of the dead animal as evidence required to collect the bounty, continued in India till 1971. The list of vermin included the jackal, each tail worth Rs 2. The Elephant Protection Act of 1879, and the earlier executive orders and Act of Tamil Nadu [Madras Presidency] banned sport hunting of elephants in order to preserve a species that was important to the colonial rulers as a natural resource of strategic importance, and not on any ethical consideration (Lahiri-Choudhury 1999). The 1879 Act was in course of time extended to Burma (Myanmar), mainly to prevent indiscriminate slaughter of elephants by British sportsmen. Elephants had by then become valuable export items from Burma to India.

The Elephant Protection Act established government monopoly in the capture of elephants in British India. Elephants could be captured or hunted only when permitted by the government. The Act recognized the destructive side of the elephant, and permitted, besides capture, the liquidation of animals for loss of human life on the one hand and of crop and property on the other.

A.J.W. Milroy (1883–1936) was Chief Conservator of Forests of Assam. Inspired probably by the 'elephant control system' practised in Africa, Milroy introduced in the late 1920s or early 1930s an 'Elephant Control Scheme' for Assam under Section 6 of the Elephant Protection Act of 1879 for reducing and keeping in check the number of marauding elephants (see Bist 2002). Such a system was also introduced by the English in Malaya (Foenander 1952). The system of proclamation for the destruction of killer or rogue elephants by a deputy commissioner or district magistrate was designed to take care of the other problem, that of rogues. It should be remembered that the price of ivory was low those days. It was around Rs 20 a seer even in the late 1960s, and rose to about Rs 40, and then to about Rs 120 at the beginning of the 1970s. This was not incentive enough for commercially motivated ivory hunting.

Milroy's scheme sought to regulate the destruction of marauding elephants. Under this scheme, only unattached adult males— that is non-breeding males—could be shot. An adult male when accompanying a herd could also be shot when the herd had been found destroying crop or property the previous day. A solitary male could be both a man-killer and a confirmed crop-raider. Rules for destroying them were separate, but with some overlap. If a crop-raider was shot, rules for awarding the trophy to the hunter were more stringent. For example, if the papers for declaring an animal a rogue were not processed in time, the animal's liquidation was recorded as that of a marauder under the elephant control licence. P.D. Stracey, a disciple of Milroy, was persuaded by his mentor to take up elephant control work under this new system. This he eventually did after acquiring a suitable weapon sometime in 1936 after the death of Milroy (Stracey 1963). The rules applied only to Assam, which in those days meant the whole of Northeast India. A licence was valid for a year and for a specified area considered especially prone to elephant damage.

The elephant control licence explicitly stated that 'it [did] not refer to elephants killed after proclamation by the Deputy Commissioner', and so, clearly, it was intended as a management tool for dealing with the problem of depredation by mature adult male elephants: 'mature male wild elephants may be shot only in areas [usually such licences were issued for specified districts] where crops and property (were) liable to be destroyed or human life endangered or in areas other than the above when found full grown and solitary' (Section 2 of the rules). Again, 'male wild elephants may not be shot when accompanying a herd unless the particular herd has been damaging crops or property within the last twenty-four hours' (Section 4, Rules). The scheme followed the principle of Section 3 of the 1879 Act which clearly put down the two sides of the problem with wild elephants: (i) manslaughter and (ii) damage to crop and property. This section of the Act states that no elephant can be killed or captured by any person unless:

(a) in defence of himself or for some other reason [i.e., in defence of life, the problem of man-killing by rogues];

(b) when such elephant is found injuring [sic] houses and cultivation ...

(c) or as permitted by a licence granted under this Act [capture].

Section 6 of the Act empowered the state [or provincial] governments to frame rules accordingly.

The ancient Sanskrit classics recognized that lone adult males were the main cause of conflict with man and responsible for most cases of manslaughter. The problem was also recognized in the 'Licence to Kill Wild Elephants' issued by Coorg in the 1920s under the Elephant Preservation Act (1879), which allowed the shooting down of mature wild male elephants. On the lines of Assam, Bengal too attempted to keep the number of adult male elephants down and prescribed a form in 1936 under Section 6 of the Elephant

Preservation Act (1879); but with the Second World War soon intervening, apparently nothing came of it.

A comparison between Milroy's scheme in Assam and the Orissa Elephant Preservation Rules (1953) is instructive. The latter was created soon after the merger of the 26 feudatory states of Orissa—the main elephant-bearing areas of the state—with the Indian Union. Milroy's scheme limited shootings to full-grown male elephants, mainly to those that were solitary, which he assumed to be non-breeding, that is, sexually retired (wrongly, on hindsight, but with the best of intentions), and tried to balance the killing of a tusker with the killing of a makna to discourage trophy hunters from going only after tuskers and thus upsetting the tusker-makna ratio in nature. With very few maknas in Orissa, this precaution was obviously thought unnecessary. The Orissa rules permitted a licensee to kill and capture [elephants] 'in any manner, except in pits', sex and age no bar (*Laws of Forest in Orissa*, 2000). In the late 1950s in Dhenkanal civil district in Orissa, the authorities ordered the wiping out of an entire herd for marauding, as they do in control or culling work in Africa. It may be recalled in this connection that only a few years ago, some Indian wildlife scientists from Bangalore recommended selective culling of adult male elephants as a management tool to contain depredation by elephants—exactly what Milroy had wanted to do. P.D. Stracey, an elephant control licensee himself, who rose to become the chief of the Assam forest department, was a recognized expert on the Asian elephant; he summed up the position in Assam in his time thus:

The scheme was devised by Milroy to *supplement* the killing of troublesome elephants [rogues]. Milroy was an astute man and knew that if he did not insist on makhnas [*sic*] being killed, sportsmen would go only after tuskers. About forty to fifty elephants are killed every year in Assam [which then meant practically the whole of northeast India] under proclamation and the control scheme, as well as a few in stockade operations, and this seems to look after troublesome male elephants fairly well (Stracey 1963).

The Wildlife (Protection) Act of 1972 radically altered this scheme of things. When the Indian [*sic*] elephant was elevated to Schedule

I of the Act in August 1977, it permitted an elephant to be destroyed under Section 11.1 (a) of the Act, as in the case of the tiger and the leopard, only when it had 'become dangerous to human life', or sick beyond recovery, but no longer for destroying crop and property. Rules and regulations made under the earlier Acts became obsolete as far as the elephant was concerned as destruction of crop and property by the elephant was no longer considered a problem requiring special attention and deserving the death sentence, even selectively.

Before 1977, the1879 Act regulated all matters related to wild elephants. Under Section 5 of the earlier Act the authority for enforcing the law was the deputy commissioner or district magistrate of a civil district. After 1972, the authority for a Schedule I animal such as the elephant, from 1977, became the chief wildlife warden of the state alone, not even another officer authorized to act on his behalf.

The post-1977 scenario does not appear to be all that happy. One cannot share Stracey's complacence that everything is under control and existing laws are looking after the problem of 'troublesome male elephants fairly well'. The main danger seems to lie in retaliatory killing of crop-raiding elephants by the affected people in the face of the government's reluctance to act. Electrocution and poisoning, which are no respecter of age and sex, are the main means of killing. In effect they are blind hits at animals, single or in groups. In the five years between 1999 and 2003 (up to 30 September 2003), 40 elephants were killed by poisoning and 169 by electrocution. (*Report of Project Elephant 2003*; the report is neither comprehensive nor conclusive). In many cases veterinary and forensic reports are permanently 'awaited' rather than available. There is a tendency in some official quarters to try to pass off unnatural deaths as natural deaths. At least seven cases in West Bengal have remained 'under investigation' for a long time for lack of these reports (Reports, West Bengal, 1999–2003). The total number of humans killed by elephants in West Bengal from 1986–7 to June 2004 was 1073. During the same period *ex gratia* relief paid for loss of human life and property was Rs 981 lakh, and 105 elephants were illegally killed. Relief for loss of life and property is generally paid *ad hoc* and *pro rata*. Karnataka paid *ex gratia* relief of Rs 2.42 crore in the five years between 1997 and 2001 (Seth 2003). The total number of human deaths in India between 1991 and 2003 (up to 30 September 2003) was 2856 (*Report of Project Elephant 2003*). The director, Project Elephant, reported to the eighth meeting of the Steering Committee in April 2002 that while the figures for poaching for ivory were decreasing, cases of killing by electrocution and poisoning were on the increase.

According to the Assam forest department, poachers killed 41 elephants from 1989 to 1997, 10 died of gunshot wounds and 36 died of electrocution and poaching, while the cause of death in 37 cases was unknown. The numbers given appear to be overlapping (Talukdar and Barman 2003). The official report perhaps does not reflect facts. The case of mass poisoning of crop-raiding elephants on the border of Nameri National Park on the north bank of the Brahmaputra in Assam in 2001 is not mentioned. According to most unofficial estimates the toll there exceeded 30 elephants.

The situation in Keonjhar, Orissa, provides a more detailed picture. Only 35 per cent of the villagers were aware of the system of *ex gratia* relief, and only 20 per cent of the people who suffered damage on record had received relief—only for the loss of paddy. There was no reliable data for the actual loss of crops to elephants and other animals. In 2000 there was a three-year backlog in clearing the sanctioned claims for relief. While villagers spent 144,00,000 man-days guarding crop in one year, the forest department staff spared merely 390 days in the field, over the five-year period 1995–2000, to assist villagers in resisting the depredation of crops. The villagers could do nothing more sophisticated to combat marauding elephant herds than shouting, beating cans, and waving flaming torches. Solitary males caused 96 per cent of the manslaughters, mostly over standing crop or when raiding stored grain.

In the official records, of the 38 cases of elephant deaths in Keonjhar between 1990 and 1991 and 1999 and 2000, 21 cases were recorded as natural deaths. An unofficial enquiry found 37 out of these 38 cases were due to illicit killing. The casualty list included 12 females, 3 calves and 12 tuskers. Interestingly, the forest department recovered the tusks in all the cases, which showed that these were not cases of commercially motivated ivory hunting.

It is, therefore, time to reinvent the wheel, to reconsider if elimination of selected animals, despite occasional lapses in identification, is preferable to allowing indiscriminate slaughter, which may result from official inaction. Modern scientific techniques can identify habitual marauders or confirmed man killers from DNA fingerprinting of bodily fluids in dung. The process, however, is perhaps not yet a practical management tool for lack of speedy and sophisticated laboratory support in the field: laboratory reports are still awaited for example, for the cause of the aberrant behaviour of a female elephant that killed more than ten people in one day in 2003 in the Kurseong forest division in North Bengal.

Human and elephant deaths and damage to crop and property are interrelated issues in the problem of man–elephant conflict: one triggers the other. Despite the expected opposition from animal rights activists, there is a desperate need to rethink the present policy, and heed the warning, 'Remove the problem elephants before people begin to remove all of them, like in Assam, or West Bengal [or Orissa] or elsewhere' (Seth 2003).

I write in this section of experiences of dealing with marauders under the old dispensation in Assam, and with killer rogues under the old as well as the current order. This experience is not limited to the act of killing, but is a package consisting of encounter with animals, men, and travel and toil. Animals that got away, or nearly did, are more important here than animals put down with one or two clinically precise shots.

In the Raj, hunters looked back on experience relishing the chase. For example, of 17 March 1891, the Maharaja of Cooch Behar notes, 'We had *khubber* both of Tiger and Buffalo. We worked for both but the tiger wasn't home, and the bull got away wounded . . . . A Rhino also got away wounded later on the day and darkness prevented us from following it up . . . . The 19th was hardly a better day though we bagged two buffalo we lost two others wounded, as well as a rhino which I knocked over but didn't, I suppose, hit in the right place' (pp. 80–1). Earlier, records the Raja of Coochbehar,

'15th March [1886] had been a better day. We bagged five Rhino before luncheon. I do not think this record has been beaten' (p. 42). The experiences I outline in the pages to follow have a different provenance. It is not the size of the bag that is the theme of my recollections but the leeches, mud, people and bad roads and impenetrable cover that my close encounters with the Indian elephant consist of.

REFERENCES

Abul-Fazl Allami, *A'in-i Akbari*. Tr. H. Blochman. Calcutta: Asiatic Society of Bengal, Series Bibliotheca Indica, 1873; 2nd rev. ed. D.C. Phillott, 1939.

Bist, S.S., ed. *Management of Elephants in Captivity*, by A.J.W. Milroy (1922). Repr. 2002. Dehradun: Nataraj.

Elison, Bernard C. *HRH the Prince of Wales's Sport in India*. Ed. Sir Henry Robinson. London: Heinemann, 1920.

Foenander, E.C. *Big Game of Malaya*. London: Batchworth, 1952.

Game Books of Gouripore, Assam. Unpublished.

Lahiri-Choudhury, D.K. *The Great Indian Elephant Book*. Delhi: OUP, 1999.

Laws of Forest in Orissa. Ed. P.K. Ray. Bhubaneswar: The Law House, 2000.

Maharaja of Cooch Behar. *Thirty Seven Years of Big Game Shooting in Cooch Behar, the Duars, and Assam*. Bombay: Times Press, 1908.

Reports to State Wildlife Advisory Committee, West Bengal, 1999–2003.

Report of Project Elephant (2003) to the Steering Committee on 22.XII, 2003.

Sanderson, G.P., *Thirteen Years Among the Wild Beasts of India* (1978). Edinburgh: John Grant, 7th ed., 1912.

Seth, Nitin (2003).' Afterwards an Eerie Silence; on elephant-human conflict in the post-Project Elephant era'. *Down to Earth*, 31 March 2003.

Stracey, P.D. *Elephant Gold*. London: Weidenfeld and Nicolson, 1963.

Smythies, E.A., *Big Game Shooting in Nepal*. Calcutta: Thacker Spink, 1942.

Talukdar, B.K. and R. Barman. 'Man–Elephant Conflict in Assam, India', (2003). *Gajah*. No. 22, July 2003.

Whitney, Casper. *Jungle Trails and Jungle People*. London: Werner Laurie, 1905.

8

~

Initiation
Cachar, October 1960

I

As unforgettable as a first love, a first kiss or a first leech sting is one's first elephant.

We were then firmly under the spell of African hunting adventures by Samuel Baker, Selous, W.D.M. Bell, John Taylor (Pondoro), John Hunter and others. Commercial films on African big-game hunting such as *Mocambo* or *King Solomon's Mines* then had a good market, as did the few Indian big-game films like *Harry Black and the Tiger*. Nobody then considered the practice of hunting morally reprehensible; for some in fact, it was a family tradition, a way of life, just as it still is with many tribal communities. Half a century ago hunting was not considered a yellow thing; for many, classical music and big-game hunting went together to shape a way of life.

I was in Shillong in 1958 or 1959 to look after some property matters. There I first met the late Mr Sawyer, the Chief Conservator of Forests of Assam, at Shillong Club. Assam had not yet been dismembered. The capital of undivided Assam was Shillong, and Mr Sawyer, a Khasi, was the head of the Assam forest department. A hearty man, he became heartier and heartier and his face got more and more flushed as the steeds of the night trotted on. In

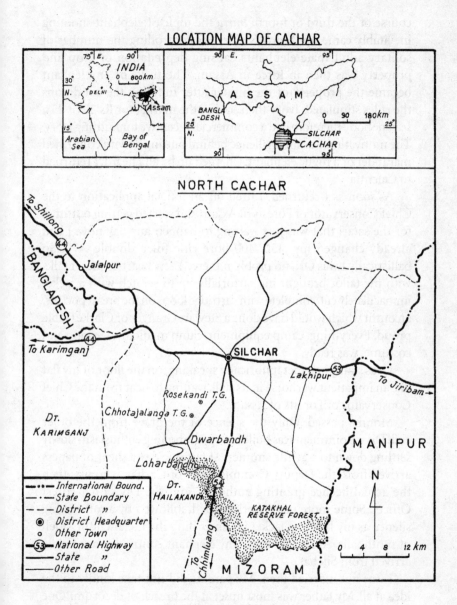

Location map of Cachar

course of the third or fourth *burra* the topic of elephant shooting inevitably came up. Milroy's rules for controlling the number of solitary adult male elephants causing depredation of crop and property was then in force in Assam. The tusks of an elephant became the property of the hunter after fulfilling the conditions the rules stipulated. Ivory then sold in the market for Rs 20 or 25 a *seer*; nobody therefore had a commercial incentive for hunting ivory. The motivation was the challenge behind putting down a confirmed marauder or a rogue elephant. Charged up by all I heard, I returned to Calcutta.

As soon as I returned, I fired off an official application to the Chief Conservator of Forests of Assam and started getting outfitted for the safari that was just waiting to happen any day now. I had already changed my .450/.400 bore rifle for a double .470. So ballistically I was OK, no problem there. A few heart-to-heart talks with my tailor brought me sartorially up to scratch with a jacket immaculately cut, complete with cartridge loops on the breast pockets, an outfit which would have done a Stewart Granger or Clarke Gable proud. Everything, camp equipment, camp medicine, weapon and costume, was ready.

I was stamping, metaphorically speaking, on the floor of my flat like an impatient warhorse in its stall, waiting to hear from the Chief Conservator of Forests of Assam.

Months passed. Only the silence of the grave from the other end. Disappointment was building up, the boiling enthusiasm slowly settling down to a gentle simmer. At last one day a sheaf of papers arrived from the Deputy Commissioner of Cachar, who was then the actual licence-granting authority, and the Divisional Forest Officer. Some more paperwork ensued, followed by another long silence as my hunting jacket lay absorbing the melancholic smell of mothballs. At last, however, my elephant control licence finally arrived from Silchar.

I was not married then. My parents did not take kindly to the idea at all. My father was most upset at the breach of decorum: 'One does not *shoot* elephants, one *keeps* them', he repeatedly announced. Mother, almost distracted, pattered from room to room after me: 'The elephant is a mighty creature. Why go after it? Why not stick to sambar, spotted deer, hog deer and the rest? Are they not enough? What's wrong with them? Why elephants? O, why, why?'

Ignoring all these admonitions with a smiling face, I caught the plane to Silchar in October 1960 just before the Puja vacation was due to start. After checking in at the Silchar dak bungalow, I went straight to the Divisional Forest Officer's office—but he was on leave and was not likely to join office before the Puja holidays were over. A kind and knowledgeable 'bara babu' (head clerk) advised me to set up camp at the Loharbandh forest rest house. This was close to the Katakhal reserve forest bordering Mizoram (then Lusai Hills) which was then the area the worst affected by elephant depredation in the district. Accordingly, the next morning I loaded my luggage onto a cycle rickshaw and went off to the bus station to catch a rickety, ramshackle bus to Loharbandh—all by myself without a second person to help me with moral and physical support. The thought of hiring a jeep locally did not even cross my mind. Did Selous or Hunter or Taylor hire a jeep? True, they did not; but it was also true that they did not start their safari with luggage piled high on a cycle-rickshaw. All these thoughts came later. I was still dreaming the African dream, of entering the forest on foot, rifle slung on my shoulder and porters following *behind* with my luggage *on* their head, in the correct nineteenth-century safari tradition. Though I was not foolish enough to dream that the people of Cachar would follow their hero, freshly arrived from Calcutta, singing 'bawana mkuba, bawana mkuba', I thought they might at least sing gratefully in a local dialect a rural tune about the arrival of the hero-would-be, the *deus ex calcutta*, to end their woes.

II

It took me a day to settle down at the Loharbandh forest rest house. I learnt from the Beat Officer that it was from this very bungalow that a manager of a local tea garden and a renowned hunter, Louis Archard, had shot two elephants only the previous year. His local tracker at Loharbandh was one Mangra Kurmi. His forefather apparently had migrated from Bihar as a tea garden labourer when Cachar had first been opened up to tea in the nineteenth century. In course of time, the clan had become completely rooted in its adopted country. Hindi, the language of Bihar, was as unfamiliar to him as it was to me. Mangra Kurmi became my factotum in the widest possible sense: at once my tracker, odd-job man, and cook, admittedly rather an indifferent one. It was from Mangra that I received my first lessons in the ways of wild elephants, their movements and tracks, their nature and behaviour. It did not take me long to realize that these lessons are not learnt from leaves of printed books, but from the leaves of trees, broken twigs and trampled grass and smears of mud on tree barks. A novice requires a teacher to explain the mysteries of the forest and read out the lessons.

The small two-roomed forest bungalow on a small hill was to be my home for the next few weeks. Below, screened off by new teak plantations, was the Beat Office, quarters of the forest staff, a few shops and the bus stand. A bus plied between Silchar and Loharbandh once a day. One could hear the hoot of the bus sending out the last call for the passengers to board, occasionally some indistinct loud voice from the shops, but actually see nothing of this tiny forest settlement. Facing the bungalow was the light green of the paddy fields and the darker green of the tropical evergreen forest beyond. From behind the bungalow stretched an unmetalled road towards the border of Mizoram. Before that the road turned right towards the river Lalamukh and at that turning was a small forest rest house at a place called Bilaipur, about seven miles from Loharbandh. (In those days we still measured distance in miles, not kilometres.) Behind the Loharbandh bungalow rolled the uninterrupted stretch of the silent evergreen forests of Cachar, extending right up to Mizoram, merging with, if memory serves, the Inner Line Reserve Forest of the Lusai Hills, now Mizoram.

Next morning Mangra and I set off for Bilaipur with two hired labourers, each carrying my belongings fastened to the two ends of a long pole of split bamboo carried across the shoulder. Seven miles was no great distance, I thought. Selous or Taylor covered fifteen or twenty miles between puffing two well-filled pipes. I had read of our own Corbett setting off in the morning, his rifle on his shoulder, climbing four or five thousand feet up a Himalayan spur and sipping a relaxing late-morning tea, oversweetened with jaggery, in a wayside village shop.

Things were not working out in the preconceived African mould. Our two porters overtook and trotted past us, leaving us *behind,* carrying our luggage hanging from the two ends of their poles, not on their head. Mangra shouted an instruction to them to halt at Bilaipur where we planned to stop and set up a temporary camp in the heart of, according to Mangra, the most elephant-infested country in the area. The unmetalled road had deep ruts made by the wheels of bullock carts. In the dry season these ruts were filled up with any kind of readily available debris, and contractors' lorries plied along the road carrying forest produce. The road went up and down small hillocks. After a couple of miles I said to myself, 'Damn Selous, Corbett and the rest', and handed over my heavy rifle to Mangra to carry. Did not the heroes in Africa use gun-bearers? Corbett possibly was too poor to afford the services of one. After years of romancing I was finding it difficult to come to terms with reality. We had started our march from Loharbandh at eight in the morning expecting to reach Bilaipur by eleven and set up camp. Our progress did not keep pace with our planning. After an hour I found that we had covered only two miles. Going forward another mile we came across a small village, where our porters were waiting for us. The village was on the right of the road to Bilaipur, the reserve forest on the left of the road. Everybody knew Mangra in that village. Just the previous year Archard, guided by Mangra, had shot a maljuria (male group) pair of elephants at night on paddy in that very village. Excited villagers surrounded us. The previous night a big solitary male had raided their paddy. After a night-long feast on the ripening paddy it had at last gone back to the forest only at the first light of dawn.

I was thrilled to see the footprints of my first elephant on my first safari. I had, of course, seen footprints of wild elephants before, but those were not legitimate game. Mangra said the elephant should be close by and we should try a spot of tracking. Asking the porters to carry our luggage to Bilaipur, we entered the forest following the spoor of our quarry. Mangra persuaded one of the villagers to join our party. This, I later realized, was the standard practice—you engaged persons with local expertise when tracking an elephant because of their intimate knowledge of the terrain. Every villager knows the forest around or close to his village like the back of his hand. He is the master of the four or five square miles of forest around his village and knows its every tree or grassy depression. Take him out of his territory, and he is completely lost. So while Mangra was my guide, the villager was Mangra's guide for that particular forest patch. Usually when you asked villagers to accompany you, they preferred to come in a party of at least two. They did not trust outsiders much when on dangerous mission; hence a companion. It was only because of the complete trust the villagers had in Mangra that a villager had agreed to come forward alone.

Within a few minutes of entering the forest, my canvas 'hunter' boots, and the lower parts of my trousers were one wet, muddy mess. There was an itch and a stinging sensation somewhere above one of my heels. Trying to scratch the place, my fingers touched something small and gooey. It was my first leech in Cachar. I had been truly blooded on my first safari. I do not remember reading of leeches in any story of African adventure. I now firmly believe that without plucking at least a thousand leeches off one's body none has the right to speak on the forests of Northeast India or its wild elephants.

The imprints of the elephant's feet were so clear on the soft ground that one did not really need an expert tracker to follow the spoor. After about half a mile we found some dung bolus on the ground. Looking at me Mangra thrust the big toe of his foot into a bolus, and with a smile whispered that it was warm: the elephant must be close. Another few cautious steps and suddenly Mangra, at the head of the party, froze, lifting up a hand in warning. Following his lead we also froze. He cupped an ear and pointed his forefinger somewhere towards the front. A few moments later I heard the sharp sound of the exhalation of the elephant's breath through its trunk, and the noise of the breaking of some branches. Within an hour of entering the forest we had caught up with the previous night's miscreant.

Later I repeatedly experienced this. After a night's feast on crops when a solitary male enters the forest cover in the wee hours of the morning it tends to seek out a resting place for the daylight hours within a mile or two of the crop fields. It does not look for dense forest cover. A large solitary male, a goonda, is a field-smart thug, a self-indulgent roughneck, a desperado that cares only for itself. It is bold without much sense of familial responsibility. It is not concerned about the safety of calves and their mothers. It is self-absorbed and is averse to unnecessary exertion. When the paddy fields will have to be visited again in the evening, why bother to go far? It lurks in any kind of cover which it finds handy. This is not true of elephant herds. They move over a larger area, and prefer staying away from human habitation during the day. I admit, however, that I have seen many exceptions to this, particularly in South Bengal.

This is perhaps the right place to describe the forests of Cachar. The Katakhal reserve forest covers a series of low hills with flat land between them. At the end of monsoon these flat tracts are covered with tall, thick grass and tara grass (*Alpinia alughus*); the hillocks have exuberant growth of *muli* bamboo (*Melocanna bambusoides*). Champion and Seth in their *Forest Types of India* categorize these forests stretching between Cachar and adjoining Bangladesh as a distinct type of evergreen forest of Northeast India. Its tall and close canopy does not allow sunlight to penetrate to the ground. Everything has a damp feel, particularly during the rains and immediately after. Narrow streams, locally called *chhara*, sometimes only a few feet wide, form a network on the ground. Elephants prefer to move along these watery lanes. Enormous ferns, seven to eight feet wide at the base and proportionately long, droop from the sides and form an arch over the stream under which the stream's water flows and the ground fauna of the forest goes about their business. Occasionally one finds a tree fallen across the stream, covered with moss and fern. At places there is a smear of mud on the rotting bark of such a fallen log, the signature of an elephant going over it. From the state of the mud, wet or dry, one can try to guess the time of its passage. Everything appears submerged in a diffuse green light. When walking along the streams, pushing through the heavy, almost oily, viscous air under the arch of ferns, one has the impression of walking across the floor of an aquarium with decorative greenery all around, light filtering down from above through a beautiful macramé of varying shades of green. The forests of Cachar are a strange, silent and eerie world. Apart from the elephant larger mammals are few. I have seen occasional footprints of wild pig, but those of deer are rare. I had heard reports that near Lalachera there was an extensive swampy ground covered with tall grass and *Alpinia*, which had quite a number of deer.

A close-canopy, evergreen forest is the natural habitat of small animals, mostly arboreal, and insects and butterflies, besides birds niched in the upper layer of the canopy. They represent the biodiversity of these forests, not the larger ungulates and predators. We find a similar description of the forests of Cachar in a late-nineteenth-century book by Edward Baker (*Sport in Bengal*, 1887) who knew the area well.

In the last forty-five years or so, elephants have nearly disappeared from Cachar. Very likely illicit felling has opened up the canopy, and encroachers have occupied some forest land. I suppose, with the habitat, the biodiversity of these forests has changed too. But leaving these grave matters for official reports and scientific journals we shall go back to the elephant we just tracked down in the forest.

Mangra's soundless lips said that the elephant was in a 'thal' (the flat land between two hills). For the life of me I could not understand how he knew that. Later I realized he must have deduced it from the squelching sound the animal was making in the mud. Mud means flat land and excludes hills; hence, it must be a thal. 'Elementary, my dear Watson!' It was from these chance lessons that I started learning from Mangra the language of the forest's many voices.

Experience taught me later that when tracking an elephant and eventually locating a resting animal, one has to adopt a certain tactic. A solitary male elephant, when resting, usually faces the direction of the track along which it had come. Therefore, straightforward spooring has to be abandoned in this final stage, and one must try an out-flanking movement. This is crucial in very dense cover with zero visibility. Many sportsmen have noted this tendency in the solitary Indian bison (*Bos gaurus*), but none that I know of in solitary male Asian elephants. This was another lesson Mangra taught me in the forest without a word being spoken—just by demonstration.

The animal was in a small patch of thal stretching over barely one-third of an acre, a hollow surrounded by low hills. A gap in the hills led to a much wider flat ground beyond. Led by Mangra I almost circled the entire thal. I could see the high spinal ridge of the animal from one spot through a gap in the green screen. It was having a great time, throwing mud and water over its back. We almost went round the whole plot but could not get a clear view of the animal, although we were never more than 25 yards away, so dense and tall was the cover of *Alpinia*. In the still air of that hill-girdled hollow, the animal could neither smell nor hear us, as the little sounds of our cautious footfall was smothered by the elephant's own noise in the wallow. Though at times in foolhardy desperation I pushed through the wall of *Alpinia* and was within ten yards of the animal, I could not see its entire body or head; I caught glimpses of the top segment of its body, the flapping of the ear on the near side, and occasionally the tip of its uplifted trunk squirting a mud-and-water douche. Needless to say, Mangra had fallen back behind me. It was now my job; his job as a tracker was over. Even a greenhorn like me realized that it would be foolish to approach any closer.

Frustrated, we came back to the outer slope of one of the surrounding hillocks. The elephant was standing just in front of us beyond the wall of *Alpinia*. We could hear every bit of sound it was making: the sound of its dung boluses hitting the muddy ground with a splash, its exhalation, the rumbling in its tummy, the flapping of its ears. We could see the tips of *Alpinia* being pulled down, but we couldn't glimpse the animal itself. The elephant was frustratingly rooted to one spot, not moving even a few steps.

Later experience taught me that this was routine with a solitary elephant, at least in Assam. In hot weather it rests from about nine or ten in the morning up to about two in the afternoon in a shaded spot, preferably with water and cooling mud. Sometimes it sleeps lying flat on one side. As the sun starts dipping to the west, it is on the move once again, resuming its feeding rhythm. This biorhythm varies in winter. The midday resting period shortens and the animal starts moving shortly after midday. A herd of elephants, however, normally does not stand still at one place in the forest for a midday rest, but moves around a lot more than the solitary male. But even herds, particularly where fodder is not plentiful, select a spot in the shade for their afternoon rest, and limit their movement to a small area. This is commonly observed in the scrubby forest patches in North Bengal and the dry forests of southwest Bengal. This makes sense. There is little that these forests have to offer as fodder. The main attraction here is crop, which can be raided only under the cover of night. Then why waste time and energy moving about in the forest in the daytime? Better to conserve energy for the night's serious work which gives adequate returns for the energy expended.

Around two in the afternoon the elephant started moving again and feeding. No such luck for me, though. All this while I had been standing next to the elephant, beating off pangs of hunger. Unlike Mangra and our friend from the village, I was not accustomed to starting the day with a hearty meal of rice. I had planned to reach Bilaipur late in the morning and have a nice, warm, cooked lunch there. The sudden appearance of the elephant midway had upset the carefully laid plan.

Eventually we surmised from the sounds that the elephant had entered the wider thal along the narrow, connecting corridor. We resumed tracking once again and entered the animal's new playground, once again covered with thick *Alpinia*. There was no way of entering the cover apart from the narrow track the elephant had created by trampling the *Alpinia* down. On two sides were solid walls of the plant, each about two inches or more thick at the base and ten to twelve feet or more tall. Even along the elephant track one had to move very carefully, putting each foot down precisely in the tub-size depressions in the mud created by the elephants' feet. In these depressions, the pressure of the elephant's feet displacing the semi-liquid mud exposed the more firm ground underneath. If one stepped elsewhere, one sank into the soft mud up to the calves. Then pulling out one's feet produced some obscene sounds which, apart from the question of decorum, were not permissible with the elephant only a matter of paces away. The elephant was proceeding slowly with frequent halts in the usual dignified pace of the larger males of the species. We were moving just behind it, carefully synchronizing the sound of our movement with the elephant's.

From time to time elephants stop suddenly for a few seconds, abruptly cutting off all sound including that of the flapping of ears, as if to make sure that no intruder is around. We too froze at these moments. One must adapt oneself to the rhythm set by the elephant. This was another lesson I learnt as I was waiting for my first brush with an elephant. We were more or less following the track and the sound ahead without, so far, actually seeing the elephant.

The elephant was not moving in a straight line but following a meandering course with frequent twists and turns in the narrow passage it was making. The flattened *Alpinia* plants pointed out the direction taken by the animal. One can tell the direction taken by a heavy animal from the direction in which trees, plants, bushes, or even tips of grass are leaning. When the nails of an elephant do not leave marks on the ground, these are the signs to look for.

Our elephant, from its sounds, was just ahead of us, but always hidden behind the next twist in the lane. At last after an hour—and about 200 to 250 yards—from our starting point we suddenly saw the rear view of an enormous elephant, at least a ten-footer. We had known from the size of its footprints that we were following a huge animal, but seeing footprints is one thing, and seeing the whole animal is quite another.

I turned my head and in a soundless whisper asked Mangra if it was a male or a female, as no glimpse of a tusk was visible from our position in the back. If I shot a female by mistake, there would be no end to the shame and ridicule I would face. I should probably lose my licence. My ignorance in this matter was profound. Everybody knows that female Asian elephants do not carry tusks. The trouble is that some males also do not. Such males are called makna. Mangra, his eyes sparkling with excitement, shot back: 'makna'. After years of a life of disputations over trivia in college canteens, I was not about to accept such a sweeping statement unquestioningly. I was yet to learn that it could never be a female elephant, a fact obvious to Mangra. A mature female elephant would never move about alone. Nor could it possibly have attained this size. I also had no idea then that the way this elephant held its head high was a sign of it being a makna.

It was a ludicrous situation. Less than ten yards behind a ten-footer animal three adult men were discussing various fine academic points about sexing it. Then ending all doubt, the animal started micturating with abandon and with the free force and flow of a cataract, displaying ocular proof of its sex. All doubts were now at rest, but technical problems still remained: how to reach from the back the vitals of the massive animal, all located in the front part of the body. African books on elephant-shooting sketch the vitals of an elephant in profile. Even Sanderson on the Indian elephant shows the passage of a brain shot in a skull drawn from the side. Stracey has photographs of an elephant's skull showing a bullet hole in the back of the head. I was more concerned here with the animal's other end. I had read a few accounts of shots at the root of the tail, paralysing and anchoring an animal, but at such close quarters I did not have the nerve to attempt it. It was obviously impossible to make a silent approach from the side through that impenetrable *Alpinia* standing in foot-high mud.

Shadows were already lengthening in the forest. There was not time enough to play the waiting game. Just then the elephant turned slightly to the right to grab some succulent *Alpinia*, and exposed its right shoulder blade. I immediately drew a bead on what appeared to be the 'behind-the-shoulder' spot, so enthusiastically recommended by the African Nimrods of the black-powder era. But I had blundered because of my inexperience and impatience. I later realized that this shot was for animals standing at right angles to the marksman. From a position even slightly behind, the recommended spot was misleading as it did not take into account the depth of the vitals from the surface which could be several feet behind the point on the surface. When aiming at the vitals, one must think three-dimensionally, not in terms of a spot on the surface. It is here that all the mystery of elephant-shooting lies. In this case, with the animal standing in a slightly quartering-away position, the angle was too shallow, and the projectile must have missed the vitals within the body by several feet of muscle, sinew and bone, and caused only a flesh wound. These are unpleasant details, which, however, are necessary to know when on control work or liquidating rogues as management prescription. The animal acknowledged the hit with a short squeal and ran straight along, smashing through the *Alpinia*, and disappeared from view. We could hear it going up the low hill on our left, breaking twigs and branches. Mangra's face was grim. There was an injured animal ahead, but daylight was fading fast.

Now this was a completely altered scenario. A novice hunter led by an unarmed local tracker and accompanied by only a villager was trying to track down an injured elephant. By no stretch of imagination could it be called a strong force.

The animal was not following a well-worn track but making a new path—up the hill through dense bamboo—that was extremely difficult to follow. The previous year had seen a gregarious flowering of the muli bamboo all over the region. The animal was bulldozing its way up through that rotting and decaying mass of bamboo on the side of the hill and the emergent new bamboo shoots coming up through that deep litter. It was tough work pushing through all that following the spoor of the animal. There was only a drop of blood on a leaf here or on the stem of a bamboo there to assure us that we were on the right track. Many of the African lessons had failed me so far, but I still held to the maxim of my African gurus: never leave a wounded animal behind. It was one's moral duty to follow it up, finish it, and spare it an agonizing slow death; so say all the 'true' shikaris', at least in print.

The light in the forest was thickening by the minute. My *bonus angelus* told me to stop as there was no longer the time or the light to go after an injured, dangerous animal; the *malus angelus* crackled: 'surely a bit more, just up the next hill, the next ridge'. In sheer pig-headed foolhardiness, at that moment I forgot another of the great maxims from the African bush: 'neither in fear nor in foolhardiness'.

After a whole weary day's wait without food and worn out by tension, I was going up and down hills mechanically, my alertness sapped by fatigue. Once again Mangra was tracking in front, the villager and I following.

The track, winding around a hillock, was going down to a flat bushy patch about fifty feet below. The western sky still glowed, but it was quite dark below where the track led. Just before starting on our descent Mangra suddenly halted and held up his hand, by now a familiar warning signal. Silently I asked him, 'What's the matter?' He replied, 'Tain' (i.e., 'tini'), a highly respectful variant in Bengali of the more casual 'he' or 'him'. I asked him how he knew it was the elephant. He said that he had heard the flapping of its ears below. I had not registered any such sound. Complete silence for a few seconds, and then a single sound of ears flapping. The sound was not repeated, nor was there any sound of the breaking of twigs or the animal feeding. The animal was on the *qui vive* and lying in ambush for its pursuers at a place of its own choosing. Without Mangra we would surely have walked into the trap.

I then decided that enough was enough, and that it would have to wait for the next day. It was getting dark, and we had deviated enormously from the Bilaipur road. Our friend, the villager, said that there was a taungya close by which could shelter us for the night. A taungya is a temporary forest settlement. Where the forest department wanted to clear a natural forest and raise a plantation of commercially valuable timber species, the labourers engaged to clear the land and raise the plantation were allowed to grow their crop for three years or so between the rows of the plantation's newly planted seedlings. In their own interest the settlers de-weeded the strips cultivated with their own crop, and in the process prevented the newly raised plantations from being choked and suppressed by weeds and undergrowth. It was a mutually beneficial arrangement. The taungya cultivators got their land free; the forest department saved the money and the trouble of getting labourers in remote areas for regular cleaning and deweeding of the new plantations. After three years, when the seedlings had attained a reasonable height and were out of the danger of being smothered by undergrowth, the taungya people were shifted to a new plantation site, and the same cycle was repeated. Shifting to a new site was beneficial to the taungya people also as in this *jhum*-type dry cultivation, the soil is very rich in nutrients for a short time, but the nutrients in the soil get washed away very soon in an area of high rainfall. This was the way of life for a good number of tribal people in Cachar then, as it still is for most hill people in Northeast India. Their permanent villages could be far away from the temporary taungya settlements.

Our friend from the village led us to one such tiny settlement—hardly a village—on a low hill quite close by. Only one Kuki family occupying two huts made up the settlement. Such families were scattered over the area clear-felled by the forest department, each family separately tending the taungya plot allotted to it by the department—something like the jhum cultivation practised by many hill tribes in Northeast India. The only difference here was that the land had been allotted by the government, not by a local tribal chief.

A Kuki family is a happy family. There were smiles all round to welcome the self-invited guests. The only glitch was that they had no food to offer to these guests. As was customary, they had finished their evening meal just before dark. Burning precious and scarce oil for something as mundane and routine as a meal was an uncalled-for luxury. They had a handful of husked paddy which Mangra and the villager boiled, made into rice, and ate with salt and a few home-grown green chillies. All the taungya people had had a community feast the previous night with contributions from the families temporarily settled there. They had brought back with them leftovers of the *pièce de resistance* of the feast: *kutta pilaff* (dog pilaff or pilau). The process of preparing this highly prized delicacy is quite simple. Take a live dog, force feed with husked paddy mixed with turmeric and salt and chilli, and when it refuses to have any more, ram the mix down its gullet with a stick, and tie up its muzzle with a thin wire or thick string, and roast it over an open charcoal fire. The result, in more sophisticated culinary terms, could be called 'stuffed dog roast'. I suspect this is another trace of the cultural impact of Southeast Asia on Northeast India's hill tribes. Despite the generosity of our hosts in offering us the best they had, Mangra and the villager declined the offer, and so did I, very politely and with profuse thanks. I opted instead for some half-ripe, tiny, semi-wild bananas they had. I noted with a wary eye that very little of the boiled drinking water we were carrying was left after a whole day's strenuous slog. The sleeping arrangements offered were quite adequate: a tribal cotton rug folded into two for a mattress on the bamboo floor and even a mosquito net; but so exhausted was I that I could sleep only fitfully.

The next morning after completing the morning 'ablutions' at a nearby chhara, watching out for leeches all the while, and after appropriating our hosts' last banana, we resumed tracking once again from the point where we had stopped the previous evening.

I can no longer remember the details of our plodding on and on that day; only the tub-size footprints always in front of us on the ground remain stamped in my memory. Our supply of drinking water gave out soon. Unhesitatingly I filled up my flask with the cool and tempting water from the chhara, ignoring the certain presence of myriad bugs in that deceptively clear water. The days of small phials of water disinfectants had not arrived yet. (Later I had to pay the price of this recklessness.) I remember Mangra pointing out a spot on the ground where our quarry had rested lying on its side. There were marks of the elephant feeding at places, but no longer any trace of blood. We spent the night in another Kuki hut. We got some flattened husked paddy there. With salt and chilli, it tasted ambrosial.

The third day began the same way: picking up the spoor where it had been left the previous evening. By then I had lost all sense of direction, and no longer had any idea which way Loharbandh or Bilaipur were. Mangra and the villager seemed equally confused. Unfamiliar with this part of the forest, they were constantly arguing among themselves. I remember we hit the old Cachar–Aizal road at one point. A strip of high land signalled the remains of what had once been a highway linking two district headquarters. Mangra did not seem too sure of his direction even after that. He pointed out the three-foot-wide track of elephants coming down along the road from the Lusai Hills to the plains of Cachar. All the grass had been trampled down. The track was thick with the marks of elephants' feet. Big trees had come up in the middle of the abandoned road. The elephant track bypassed them. At places trees had fallen across the road. I had my tiffin of flattened husked paddy sitting on the trunk of one of the fallen giants of the forest, relieved at seeing the blue sky once more over my head.

Looking back, my memories are smothered in a swab of wet green. Seeing this forest from outside and from within are two different things. From outside it is a massive peristyle mansion stretching from horizon to horizon, edged with tall boles of trees like pillars; seen from the dungeon within, it is a completely different world.

Its inhabitants live their own secret life, move along paths and lanes of their choice, shelter in the mansion's hidden chambers and cells when pressed by intruders from the world without. The first day I had felt that everything was floating in green light. By the third day it seemed as if everything was soaked in green ink which would stain my very skin. We, the creatures of light, may be charmed and fascinated, viewing the green mansion from outside as an ornament on the landscape, may even enjoy being within one of its halls; but after a while one longs for light, and longs for a deep breath of crisp, clean, sun-drenched air. Like some trees, we too demand light—without it we cannot survive.

The second day's adventure with chhara water was proving costly. Every now and then I had to squat behind some cover, ignoring the threat of leeches, to pay the price of my indiscretion. My feet had swelled after three continuous days of plodding on animal track winding up and down hills. The sand my canvas hunting shoes collected from the chharas as I waded through them had serrated their heels, sandpapering the back of my ankles to a raw sore. I could walk only because Mangra cut off with the villager's *dao* the upper portion of the shoes' heels—a piece of inspired surgery.

We had lost the spoor of our quarry at some point, and were trudging mechanically along the chharas. Mangra offered the redundant explanation that we were lost. We had no idea of the direction of either Bilaipur or Loharbandh. Mangra advised that we should follow the flow of a chhara. The forest was on high ground surrounded by villages and cultivation in the valleys. The chhara water would flow out of the forest to a valley. This was certainly a better plan than moving in a directionless manner through a forest which stretched to the Lusai Hills. In my condition I could never have reasoned this out. I realized that I was beyond the limits of my endurance and was running high temperature.

One could no longer see in the gathering gloom the clear outline of the tops of trees overhead. Once we heard the shrill trumpet of alarm of an elephant next to a chhara we were following. Another time a small herd stampeded out of our way hearing us approach. The monotonous drumbeat of a single insistent thought accompanied each of my steps: When would this end? And then suddenly, after negotiating a bend in the chhara, an expanse of paddy fields greeted us. We were out of the green chateau through one of its water gates.

I flopped down on an earthwork divide between paddy fields. I could see village lights in the distance. Mangra spotted the place at once. Walking in circles for three consecutive days had taken us right across the forest to the other side. The place was about two miles and a half from Loharbandh. All my belongings, including my bedroll, were in Bilaipur; but Bilaipur was not an option for us just then. I can no longer recall how I made the couple of miles back to Loharbandh. I only remember crumpling down at the edge of the village road every few hundred yards. We eventually reached Loharbandh quite late in the evening and I collapsed on the coir mattress of a bare bed. The kindly Beat Officer sent me a bowl of warm milk and some puffed rice. There was neither the time nor the necessary provisions for a cooked meal that night. It took me three days to recover from the exhaustion and fever.

This experience taught me the hard way the importance of not muffing one's shot and what following up a wounded animal entailed. I never made such a mistake again, except once, but it had extenuating circumstances.

The makna, like a piece of monsoon cloud, melted away in the dark forests of Katakhal. I never saw it again, but its huge footprints which I had come to know so well over three memorable days remain permanently etched on my memory.

III

For about a week I just sprawled on the spacious easy chair on the veranda of the Loharbandh bungalow, the thoughts of the earlier encounter swirling through my mind, thinking miserably over and over again about my mistakes, and the reasons for them. At the end of the week I received a message from the manager of the Rosekandi tea garden informing me that a small herd of elephants had made the garden its home and all day-to-day work in the garden was about to stop because of this herd. The deputy commissioner, after repeated complaints, had ordered the destruction of one male elephant from the herd and had asked Mr Choudhury, the manager of the garden, to contact me at the Loharbandh forest rest house.

A jeep from the garden picked me and Mangra up. Mr Choudhury showed me the deputy commissioner's order. (Before the Wildlife [Preservation] Act, 1972, the deputy commissioner was empowered to issue such orders or special permits. These orders did not come under the purview of the elephant control licence.)

Next morning along with Mangra I started off in a jeep for the Sahapur division of the garden, six or seven miles away, where the troubles were centred. The assistant manager in charge of the division, Mr Brahmachari, told us his problems. Just in front of his office, a garden road went straight ahead, tea bushes on both sides, ending at a forest-covered hillock that looked like an island floating in a sea of tea bushes. I discovered later that it had a narrow link with the forests of the neighbouring tea garden, Chhota Jalenga. It was the elephants' resting place during the day. Every evening the elephants came out of this shelter, went on night-long rampages, breaking workers' houses and destroying all their privately raised crops. Along with two labourers from the garden, Mangra and I climbed back into the jeep and drove towards the hillock. There were tea garden roads all around the hillock. It took us not even half an hour to go round it. Their complaints about elephant depredation were obviously genuine. Elephant dung boluses littered the roads, some old, some absolutely fresh. The walls of many houses in the labourer lines were broken or damaged, and the small crops the labourers had tried to raise in the plots of land around their houses were trampled over and damaged. Some trees in the plantation had been pulled down. The wire fences around it were damaged every few feet.

The labourers pleaded that after fighting the herd night after night it was not possible to cope with the work of the garden in the morning. Some members of the labourers' families had also been trampled to death by the elephants. They had reached the limits of their endurance. In the darkness of the night, it had not been possible for them to identify any particular elephant guilty of all these misdeeds; but all the alleged miscreants sheltered in the cover on the hillock during the day.

After the uninterrupted green of the Katakhal reserve, it appeared to be child's play. It was only a patch of forest, not too dense, completely encircled by tea bushes. Elephants could not possibly get out of the patch unseen during the day across the low tea bushes. Two labourers from the garden showed us the precise point in the forest, where the elephants, having helped themselves to the labour line's home-brewed liquor, paddy, and other crops in the fields, had entered the forest cover that morning.

Mangra, I, and two garden labourers—a party of four—entered the forest along that track. Following our now-established routine, Mangra led the party; I followed and behind me were the two labourers. In the latter's charge was our dry lunch in a jute shopping bag: flattened husked rice mixed with jaggery, some excellent bananas and drinking water in a big drum made of zinc sheet and covered with thick felt. This once belonged to my great-grandfather, and therefore had some sentimental value for me. The drum had served him well in many a howdah shikar in open grassland. After our recent experiences in Katakhal, we were painfully aware of the importance of having sufficient potable water with us.

Reaching the hillock, we found that it was criss-crossed with a network of elephant tracks. It was clear from this that elephants had inhabited this patch for quite some time. As we advanced along a track crowded with fresh, recent, and old spoor, all mixed up, we soon lost the fresh track we were following. Though from a distance it appeared to be a single massif, after reaching it we realized that it was actually a cluster of low hillocks forming a bowl, at the bottom of which lay a low flat patch of irregular shape forming a thal covered with *Alpinia* and soft mud. We moved from one hillock to another round the rim of the bowl without being able to locate the herd. We had gone from hillock to hillock circling almost the whole bowl when we heard the sound of a breaking branch below.

As we prepared to stalk the herd, some cold common sense washed over me. My great-grandfather's water drum was obviously perfect for a howdah, but to be stealthy and silent in its company was impossible. When we moved, the water in the metal container splashed about with a merry sound. Despite the claims of nostalgia, we left the drum on the hilly rim in the care of the two garden labourers and took an old, much-used track to go down to the pit of the bowl below. The ground there was not too soft or muddy, probably because it got more sunshine than the thals in Katakhal. As had become usual in such a situation, I was in front with Mangra closely behind me. We were moving slowly, all faculties alert, along the floor of the bowl, which had an irregular contour because the ridges from the hills coming down to it. These ridges also provided unexpected hideouts to the elephants. At one point a few yards ahead of us the track split like the upper part of the letter Y. I stood there pensively wondering which arm of the fork I should take, left or right. Twisting my neck I was about to whisper to Mangra for his opinion when I saw the whole group about twenty yards away coming straight towards us in a single file along the right arm of the fork.

We were undoubtedly in an uncomfortable position. Not knowing what I should do, I took the best possible course—I just froze, and extending my arm back, waved Mangra down; he immediately dropped down on his haunches. I silently pushed forward the safety catch of my rifle. The group was coming straight on, a large female in front. Of course it would be a grave offence if I shot the cow, a serious breach of the rules, but in a crisis, in self-defence, everything was permissible. The elephants were coming towards us in their usual unhurried, dignified pace. My forefinger was tightening on the front trigger when at the junction of the two arms of the Y the herd made a U turn and took up the left arm of the Y, offering us a grandstand view of their march past.

Even at that close a distance, not more than ten yards away, the elephants had not detected our presence: in that hollow without any drift in the wind, they could not smell us. They could not see us either simply *because* we were so close, and absolutely still. Because of the peculiar position of the elephant's eyes on two sides of its massive cranial structure with a central bump intervening between the eyes, the elephant has no binocular vision at close quarters, and cannot see what lies straight ahead when the object is very near. It is time to be wary when it suddenly stops, turns its head slightly to one side, and tries to focus with one eye. I could clearly see and count every animal in the group—there were seven of them—but I was unable to detect a makna (there was no question of missing a tusker in the group, it would have been clearly visible). There should have been a makna. But the Katakhal makna had spoilt me. I was thinking only big—too big. I missed noticing the small makna in the group.

After another broadside view, however, I spotted it and could bring the proceedings to the officially desired conclusion. The end seemed rather tame and anti-climatic after Katakhal.

Looking back, I feel the decision by the authorities to eliminate one male elephant from the herd was ill-advised and unnecessary. The herd could have been chased away from its refuge, the crater-like depression in the hilly forest patch, by just creating a disturbance there: by using koonkis to chase them around during the day or by simply lobbing crackers into the bowl from its rim which would certainly have induced the small herd to seek another haven after dark.

1. Bajra Prasad in full regalia (from the pilkhana of the Late Raja Jagat Kishore Acharya Choudhury of Muktagachha, Mymensingh, Bangladesh).
2. My son, Deep Kanta, atop Lalji's Tirath Singh. (Plates 1&2: Ch. 1.)

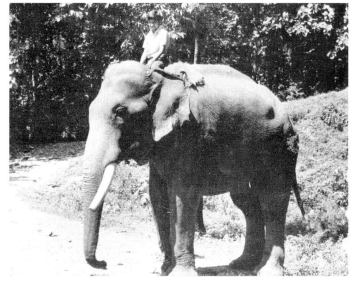

3. My wife, Sheila, feeding laddus to Maloti at Shulkapara (Ch. 3)

4. Harjit contemplating life among the tea bushes of Simulbari tea garden in North Bengal. (Ch. 6.)

7. The mahout had barely brought Jaymala back to her stall when Premnath joined her, drawn by her throaty, come hither chortles. The separation had aroused his passions, and he swiftly went about his business.

Then 'calm of mind, all passion spent'.

9. Lalji before his rowti (single-fly-tent) at Sonpur. (Ch 2.)

). Gabbar Singh in Dalma. Note the broken tip of the right tusk exposing the nerve cavity. (Ch. 5.)

.Elephant-caused devastation in Bandapani Khas, North Bengal. (Ch 14.)

12. Team of khaaddis led by Palnii going into the Bamadehii forests in 1980 in search of the same latex Prasad is just out of the frame. (Plates 12-17, Ch.15.)

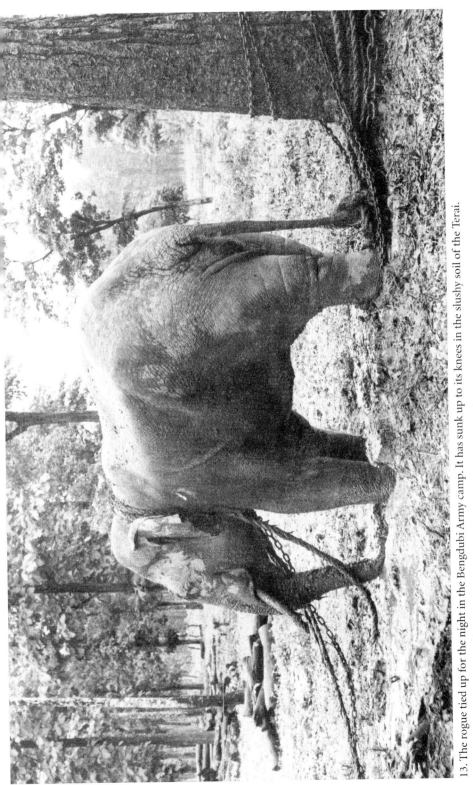

13. The rogue tied up for the night in the Bengdubi Army camp. It has sunk up to its knees in the slushy soil of the Terai.

14. Next morning it was loaded on to a lorry, pressed between Jatra Prasad (right) and Balaji.

15. After revival a captive in Malaysia is tied up in an Assamese-style noose, in which the lead between the captive and the koonkis is short. The larger animal is Fulbahadur, a trained makna. The fibre of the hand-made jute rope is from Bangladesh. In the Karnataka practice the rope is hemp and the lead is very long.

16. The captive at Bengdubi is being transported by lorry.

17. A wild elephant in Malaysia journeys by raft.

18. A wild tusker in musth moving through a moist deciduous forest.

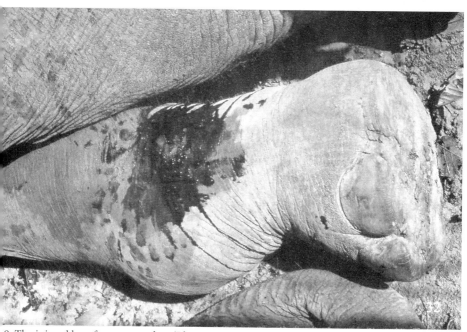

9. The injured leg of a young tusker. (Plates 19-20: Ch. 18.)

10. The leg, after being dressed. The white bandage is only for photographic effect: it would not last minute after the animal was revived.

21. The tusker that had rolled down the embankment of the Kangsabati irrigation canal taking with it, avalanch-like, half the paved lining of the canal. (Ch. 20.)

22. The tranquilized chakna on the ground is being swiftly tied to Jatra Prasad and Chandrachud before it revives. (Plates 22-4: Ch 22.)

23. The captive is led along the Jhargram highway between Jatra Prasad (left) and Chandrachud (right). The captive seems to prefer Chandrachud's company.

24. The carcase separated into the several portions which are of use to the men and elephants

9

Bloodletting
Cachar, October 1961

I

Wise from the previous year's experience, I abandoned the idea of an old-fashioned Africa-style foot safari, acquired an army-reject jeep, and got ready for another expedition to Cachar during the autumn vacation—a season traditionally preferred by the royalty in ancient India for setting out to conquer new lands, *digvijay*. Quite a crew accompanied me this time. Apart from the driver Ahmed and a cook, I had four young but enthusiastic sightseers with me this time, livewire Ashoke among them. That they were incapable of telling sal from eucalyptus did not seem particularly important then. I wanted some young friends to share my holiday spirit. I was also just recovering from a bout of infective hepatitis, and my mother thought there was safety in numbers.

The journey to Cachar by jeep turned out to be as challenging as the forests of Cachar. Floods had cut off the direct road link between North Bengal and Calcutta. We had to make a detour of more than 200 miles through the Rajmahal hills and Dumka, but the floods relentlessly followed us there as well. The jetty normally used to load cars onto the boat had gone under water. We had to wait for about two nights for the water to recede. Then we crossed the still-swollen river and somehow managed to reach Maldah. There we were stuck again for the ferry at the Mahanadi river. The rear wheels of a loaded lorry just in front of us had slipped off the boarding planks into the river: the front wheels of the vehicle were on the boat, the rear wheels in the shallow waters, the lorry looking skyward like an ancient Indian sun-gazing sage performing a religious rite. It took the whole day to unload the lorry, manhandle the lorry onto the boat and reload it. By the time we completed the river crossing it was almost dark. Then to top it all, the engine of the jeep decided to die on us. It took more than an hour's cajoling to resuscitate it back to coughing life. Our last major hurdle was at Mathabhanga in Cooch Behar, a detour the floods again compelled us to make. The powerful Willys engine of the army-model jeep with its four-wheel drive saw us through the knee-deep mud of a village lane to the river bank.

And so, limping and halting, at last we reached Shillong. After a day's rest we took Jowai Road to go across the Jainti Hills and descend to the plains of the Barak Valley in Cachar negotiating one more ferry on the way. Jowai Road was still under construction then and was in a very rough state. Huge blocks of stone jutted onto the side of the road making driving hazardous and excruciatingly slow. Every ten miles there was an army checkpost. The road was very narrow, not yet blacktop, and the traffic one-way between roadblocks. The registration number of every vehicle passing through had to be taken down. The army had set up field telephones between roadblocks with the object of ensuring that before the gate for traffic going up was opened, every vehicle of the traffic coming down at the previous gate had come through; it was a necessary precaution: a stalled vehicle on that stretch of road would have led to a traffic jam all along the road from Shillong to Cachar.

At Sonarpur, we had to take the last ferry on our long jeep journey from Calcutta to Cachar across a fast-flowing mountain stream hissing and gurgling through a gorge. The stream was a torrent after even moderately heavy rainfall. There was a steel cable anchored to stout trees and strung across the gorge. The boat to ferry us across was attached by another cable to the one overhead. A diesel motor propelled the boat across. The cables were meant to ensure that the force of water did not wash the boat away. Every few miles there were notices on the road announcing: 'Elephants have the right of way'. It was nearly evening when we reached the plains of Barak Valley on the outskirts of Silchar. It had taken us nine days of strenuous driving to reach Cachar from Calcutta, negotiating eight ferries on the way. At last Loharbandh! My home away from home!

Mangra came to meet me the next morning. After a 'recce' of the surrounding areas for a few days, I called on my old friend, Mr Choudhury of the Rosekandi tea garden. He was as expansive and welcoming as ever. Again, the centre of his elephant troubles, the most critical this year in his long experience, was the Sahapur division of his garden. I drove to the Sahapur office and was immediately surrounded by the local staff and the garden labourers who knew me from the year before. They all started speaking of their troubles at once—one problem but many voices. Had I asked them to speak one by one, they would certainly have broken into a chorus of silence. Slowly the matter became clear. A large tusker was raiding their crops and quarters every night. Even the previous night it had raided the labourers' crop. Whose paddy? Sufal's. Where is Sufal? The crowd catapulted a stocky, well-built man forward. Since it was Sufal's paddy that had been plundered, who but Sufal should show me the elephant! Popular opinion loudly endorsed the view. Sufal, carrying a long-handled Assamese dao in his hand, was instantly ready. He picked up a friend, Sudam, another garden labourer and fellow sufferer, as his companion—as was customary. We picked up the track quickly enough and entered the forest in a single file and started spooring: first Sufal and Sudam, then Mangra and I, and then the rest.

The previous night it had rained heavily. The footprints were very clear, but the mud was so soft that our movement was slow. Some time later as we were pushing through mud and dense vegetation along the narrow lane created by the elephant's passage, there was suddenly the sound of the breaking of a tree branch. We stopped at once. Seconds later we heard the long-drawn-out sound of bamboo being pulled down. Now we would begin to stalk. I readied myself by emptying my pockets of all things that might rattle. Leaving the rest of the party at the base of a large sirish (*Albizzia* sp) tree, Mangra and I were about to tiptoe forward to the right towards the source of the noise only twenty yards or so away, when there was another sound of the breaking of branches about a hundred yards to the left. The plot was thickening—our quarry had been joined by another elephant, certainly another solitary male, and had become maljuria (one of a bachelor group). After a few moments of indecision, I decided to go to the right, as the track we were following led that way and the sound was closer. There was now no doubt that there were two elephants, one very close, the other a bit farther away. Mangra and I were moving very slowly. One can never move in total silence through such heavy cover. Other sounds from the elephant were now audible: the flapping of ears and heavy breathing. Remembering the experience in Katakhal, I was trying to move coinciding the sounds of my movement with the elephant's own sounds, freezing when the elephant went still for a few moments. At last after going round a bush, I suddenly saw a large tusker standing broadside on in a chhara a few yards away. The tusks were short but thick, the body large and impressive. I thought Lady Luck was with me that day: it was most unusual in the forests of Cachar to get a clear view of an elephant as much as twenty five yards away.

Here began my series of mistakes.

The virtues of heavy and light rifles are a much-debated topic. It is agreed that a solid bullet from a light rifle, if placed with absolute accuracy, can be deadly; but pinpoint accuracy is the precondition here, which one does not need for a through-the-shoulder heart shot. This is a shot for a heavy rifle—which I was carrying.

I do not know what flashed through my mind. With the elephant's body fully exposed, I thought I should try the famous ear-hole shot to reach the brain. Sanderson called brain shots the poetry of elephant hunting and heart shots the prose. I still cannot explain why I had this sudden urge to go poetic. I shifted a bit to get the perfect angle for an earshot. After the fiasco of the Katakhal makna, I was very conscious of this point. I must have made some small noise or maybe there was a slight drift in the wind, for the elephant went still, and then slowly turned to face me. All my book learning had taught me to avoid the frontal brain shot as being too chancy. But the tusker then started coming straight towards me at a brisk pace. It was not a charge but an inquisitive look-see to search out the source of the disturbance. I was compelled to take the frontal brain shot at a distance of ten or fifteen yards, and, predictably, muffed it. I must have misjudged the angle, or the elevation, as the animal was coming up from the low floor of the chhara with its head somewhat lifted. It acknowledged the hit with a wince, and then turned sharply and ran along the chhara. I discharged my left barrel in the general direction of its receding shoulder. Again, as in Katakhal, light was becoming the worrying factor. The realization dawned on me that I had done something very foolish in bringing with me four absolutely raw young boys, without even minimal bushcraft. This was not a jaywalk or a shikar holiday, but much more serious. It was already past two. I had at least my rifle; they were completely defenceless. Obviously, I could not ask them to go back to the jeep through the forest by themselves. In another half-hour, there would not be enough light in that dense cover. It appeared that my quarry was in a very bad shape. Once again, Mangra was leading as the tracker, a repetition of the Katakhal episode. This time, following us were not only the two garden labourers but also my four young and enthusiastic city-bred friends. Soon enough we saw some frothy blood on the ground, which could have signified a lung shot. But I was not wholly satisfied. Going by the books, there should have been much more of it.

Tracking now became a serious problem. We had to go through dense grass and soft mud. After going like this for half an hour, my hope of catching up with the elephant was receding fast. After high excitement, depression was slowly setting in.

We were proceeding very, very slowly. At a short distance to our left we heard a rumbling sound. At first I thought it was our elephant. The sound was so strong and so peculiar that Mangra at once contradicted me and said that we had by now come close to the neighbouring Chhotajalenga tea garden. It must be the sound of the diesel engine of one of their lorries. I had to admit that the sound was very much like the sporadic revving up of a diesel engine, missing in one cylinder. After a while, we realized that it was nothing of the sort. It was our elephant all right. Standing on a small hillock, it was making such a din that it could be heard a couple of hundred yards away.

Closing up, we could hear the sound of another elephant, undoubtedly the elephant we had left when we originally starting our stalking. It had come to its friend's rescue in the latter's bad times. It was already well past two in the afternoon. We had only about forty-five minutes of tracking time left. Lengthening shadows in the forest were already affecting the light. The jeep was about eight miles away by the garden roads, as Mangra informed us. We needed to start getting out of the forests soon.

With Mangra, I inched forward as silently as possible. I could occasionally see the entire body of the animal now from the back. The elephant was retching and making that odd sound. I tried to approach it from the side, but hearing me it shifted a short distance. It could not go very far. The sound was coming just from a couple of hundred yards away. This happened several times. As I came closer the animal shifted its position. Through a dense network of twigs and branches, I could see a big makna pushing the tusker from behind. The same game of hide-and-seek was repeated several times. By then we had less than fifteen minutes of tracking time left before we had to begin wending our way back to the jeep at Sahapur. There would not be adequate light after that. There was no way we could go back to the jeep by the shorter route through the forest—it was already too late for that.

I pushed forward along the track, feeling miserable about the suffering I had caused the animal, and the guilty feeling of having botched it. We would now have to take a short cut to the Chhotajalenga office, and arrange to return to the jeep left at Sahapur. But the elephant was so close! I could hear every sound it was making. I did not have the heart to leave it to its suffering. I was going down a narrow animal track along the ridge of a hill when I heard the elephant rumbling in a gulley below. Feeling reasonably safe having located the animal, I was about to climb down to the gulley. The track was narrow at that point, with a thick palisade of trees and bamboo lining the sides. In the failing light one could clearly see the bare, twilit strip of the narrow track. I focused on the sound coming from below. The sky behind was a spectacular golden red such as only an autumn sunset can present. Suddenly while negotiating a bend in the track, I saw a dark, triangular shape stretched across the narrow path: unmistakably an elephant's ear silhouetted against the brilliant background. I froze. Mangra dropped down on his haunches just behind me. Within a few seconds, the shape of a big elephant's head without tusks came into focus. It was facing us straight from a distance of less than ten yards. Not a sound, the ears flared out and absolutely still, the trunk lifted up sniffing for tainted air. It was a perfect ambush. Distracted by the noise from below, I'd never have detected the animal but for the shadowy shape of its ear. Weary of being pushed by the pursuers, the makna had decided to take its stand at a place of its choice, the sound of the elephant below providing it with the best possible diversion.

Here was a first-class impasse. I already had an injured elephant on hand, the result of a muffed frontal shot. Now I again had a frontal shot nearly forced on me. I shuddered at the idea of two injured elephants in hand instead of one. I decided to remain absolutely still, my rifle up and ready, and carry out a waiting game. Another step forward and I'd have to take the much-hated frontal shot anyway, whatever the consequences.

Time seemed to have congealed. Eye-flies were clouding round my eyes; perspiration was trickling down on my glasses and I could feel a leech settling down to a feast in the area around my ankle, but did not even dare to blink. After waiting like this for an infinity, which was actually perhaps not more then a few excruciating minutes, the makna slowly turned around and went down the track to its mate, as if to warn me: thus far and no farther.

The message was well taken. We started to work our way out of the forest. The exercise took an hour and a half, and we emerged at a small Khasi village on the outskirts of the Chhotajalenga tea garden. Sudam and Sufal went to the tea garden's factory and office premises with a letter from me. From there they borrowed a bicycle to get to Sahapur and brought our jeep back to us.

The next day we took up spooring from the point where we had left off the previous day. It was like the rewinding of a cinema spool. We reversed on the same track back to our original starting point:

the hilly forest patch at Sahapur. The previous day's track carried the imprints of our cleated shoes. Fresh tracks of the tusker, super-imposed on the earlier tracks, were now pointing us to the opposite direction. By then we were familiar with the size of the footprints of the two animals, and could easily tell between them. Midway the makna had separated from the tusker. This suggested that the tusker was no longer in extreme distress, and needed no further help from its buddy. There were signs of the tusker feeding on its way back. Flattened undergrowth at one point showed where the animal had obviously lain down and rested. I should therefore be prepared to meet an animal no longer handicapped but fresh and fighting. When we finally emerged from the patch of forest at Sahapur, it was nearly dark. Another hard day's work, supported, mercifully, by drinking water and dry tiffin!

We met Mr Brahmachari who had been waiting anxiously for us at the Sahapur office. He was worried about us as he had had a nasty experience on the road from the forest to his office only a few weeks before. He was returning late from an inspection of the garden when he heard a loud hissing sound. Instinctively he raised his umbrella, only to receive a hard knock on its curved handle, hard enough to make him stagger back a few steps. In the fading light he saw a huge hamadryad slithering away into the tea bushes. As proof he showed me the umbrella: it had a big crack in the handle, almost splintering it. The strike must have had enormous force. Half an inch this way or that, and he would have been a dead man. Tourniquets are obviously impossible in the chest area, where he would have been struck—and in any case, who has ever heard of a tourniquet saving a life after a strike by a hamadryad!

The third day we again entered the Sahapur tea garden forest. The luxury of being able to say, 'it got away' was not for me. I was duty bound to clear up the mess I had created. After a short while we could hear the elephant, presumably *our* elephant, in a clump of bamboo and feeding, which meant an elephant in sound health. I could not prevent my enthusiastic young friends from accompanying me; only Ashoke decided to stay back near the jeep. He declared he had had enough of this business of elephant control, and would restrict his adventurous spirit to catching some butterflies instead. He already had a contraption made of cane, looking something like a tennis racket with a torn piece of loincloth attached to it, and was determined to try his prowess with it.

The elephant was in good health and very alert. We approached it warily a couple of times but it slipped away detecting our approach. We had the whole day ahead of us to track it down in that small patch of forest, so I was not worried. Again we heard the elephant quite close on a small hillock about 150 yards away. We were on another hillock, a narrow ridge connecting the two points, an old game track running along the top of the ridge.

Had it been a large forest like Katakhal, I would never have seen this animal again. Here in this patch of forest the animal would not break out and make its way through open ground covered with low tea bushes. Mangra and I crept along the track, which appeared not to have been used in recent times from the way it was overgrown with creepers and low shrubs. A thick growth of cane covered the two steep sides of the ridge. Ahead of us was a small hilltop covered with bamboo. Approaching close, we could make out the form of the elephant feeding on bamboo on the other side of a clump. At one point we could see just the corner of its head and left eye; but the rest of the body was hidden by dense bamboo.

The track we were following went round the clump. There was no question of firing through the dense bamboo. I could not risk the bullet being deflected. Suddenly the elephant stopped flapping its ears, as elephants do every now and then. We also stopped. I had been standing on a dry stick of bamboo when I froze. Under my pressure the twig snapped with a slight sound, and then the inevitable happened.

This time the elephant did not try to escape but moved straight towards me pushing through the clump of bamboo. We could see nothing clearly except that a large mass was coming towards us smashing through the bamboo. Bamboos were splintering making sounds like pistol shots. I had been standing as close as possible to a clump of bamboo, about two yards away, for a clear view of the animal. Now it meant that I would see it only when it broke out of the bamboo at a distance of a couple of yards. Without any military training, I had not realized the importance of a clear field of fire in front. My first impulse was to run. But there was no possible escape route down the slopes of the ridge for a mere man, so dense was the cane with inch-long curved thorns. If I ran back along the track there was every likelihood of my stumbling over a creeper and falling flat on my face and being trampled by the elephant without even any malicious intent. It had no other way to go either. I had no choice but to face the animal trusting in the power of my heavy-bore rifle. I was no hero but it looked as if circumstances would thrust heroism upon me.

In a second the huge head of the elephant burst through the bamboo smashing all obstruction. The foreshortened thick short tusks were looking even shorter because the animal was holding its head up and I was on the ground just in front of it. With its trunk curled up and ears flat against the head, it seemed as if the head was floating at an angle just in front of me at the height of the ceiling of an ordinary room. There could not have been any worse angle for a frontal brain shot. I fired straight up. A steady, well-calculated shot was impossible. The objective was to turn the elephant back, not kill it outright. The 500-grain bullet of my .470 did not let me down; if it had, I would not be writing this story. The animal staggered back a step, and like an avalanche, hurtled down the steep slope of the ridge. At once the rumbling sound of the first day started. The first day's wound, which I later found out had been at the base of the trunk, must have opened up again by the impact of it dropping down the slope.

A grim determination had by then gripped me. For three days I had been trailing this animal. How much longer? I went immediately down the slope following the passage made by the animal. Below was a thal full of mud and reeds. The animal was about fifty yards ahead of me making that loud sound all the time. If it had turned round silently like the makna on the previous day, it would have been a dangerous situation. Fortunately the makna was not there this time and the presence of the animal was clearly evident from the noise it was making. The one thing going through my mind was that I must not let it escape again. The elephant was going forward slowly, stopping from time to time and I stopped too, synchronizing my movements with the elephant's, making no attempt to close up. I had to catch it on ground more favourable to me. At some point the elephant was bound to get to a higher ground, and then would come my chance. Meanwhile I had to be on the track of the animal like a relentless wild dog trailing its prey, and on no account lose contact with the elephant. It was not going to be one of those that 'got away'.

Moving very slowly the tusker climbed a small hillock barely fifty feet high without any tree to speak of, its sides thickly covered with wild banana draped with a lattice-work of creepers. From the base of the hill one could see nothing of the elephant. Peeping through the screen of creepers, at one point about halfway up the hill, I could see the spinal ridge of the elephant. By that time the very idea of saving cartridges had become anathema to me, and I at once fired at the spine, hoping to anchor it for good: admittedly not very sporting, but perhaps unavoidable, as things had turned out to be. There was no tree nearby against which I could steady my rifle and I had not yet recovered my breath after wading through the muddy thal. I must have missed the spine by a fraction of an inch or so, as I later realized. The elephant fell with a loud thud followed by the sound of breaking twigs and shrubs. The rumbling sound was still coming from above. But with a spinal injury the animal would be firmly immobilized. I had only to go up and finish the job.

I started climbing slowly up along an old game path. About half way up Mangra whispered from behind, '"Sir" is up'. 'How did you know?' I whispered back. 'I heard "Sir's" ears flap,' Mangra answered.

Then in desperation and sheer pig-headedness I committed a serious breach of the rules of stalking. You do not stalk *up* to a dangerous animal; you stalk *down* or try to meet it on level ground. In this case it was not just *any* dangerous animal but an injured tusker, no less. I should have tried to close in on the animal by climbing up from another side, not from a point almost straight below the animal. I could see nothing of the animal through the thick screen of creepers. From behind a clump of banana I peered up to see if the animal was in sight.

Just then, the tusker charged straight down at the clump of banana behind which we were sheltering. I do not remember hearing any vocalization by the animal at the moment of the charge. I could only judge from the sound of shaking and tearing down of bushes that something large was rushing down on us. There was nothing I could do as I could not even see the animal; even if I had been able to, an effective shot could bring its three- or four-ton body rolling down, crushing us. I instinctively tried to step back, slid down for a few feet, and then remained suspended caught in a thorny bush, still bravely looking towards the oncoming wave of sound. Petrified, I saw a trunk, thick and rough like the bole a young sal tree, and a foot as massive as a pillar smash the clump of banana behind which we had been standing moments ago. It is still a wonder to me how the animal could pinpoint our presence about thirty feet away down a hill slope with such deadly accuracy, just by some slight sound we might have made. In my thorny cradle about ten feet further down I was still managing to hold on to my rifle and staring in the direction of the danger. For a few minutes I remained there absolutely still. The elephant did not charge further down. Its rumbling noise was coming from above. I was terrified that any sound I might make, however slight, might provoke another charge. There was no sign of Mangra.

Keeping my rifle ready, I shook a small twig. This did not bring on the tusker. My bullets were perhaps taking effect slowly. I tried the gambit a few more times, shaking a twig. There was no answering movement from the tusker. Courage slowly dripped back into my veins. I softly whistled through my lips for Mangra. A soft whistle responded to my call from somewhere about twenty feet further down. Cautiously I freed myself from the clutches of the bush and started a crab-like crawl downward, keeping my eyes focused ahead, careful not to show any disrespect by turning my back on the royal presence of the tusker.

When I eventually caught up with Mangra I was impressed by his sangfroid: with a wry smile he said, '"Sir" gave us a good chase.' I had a feeling that in his eyes I was slowly coming out of my novitiate. The tusker had notified us clearly that it was not immobilized or dead. A message also came through that it was nevertheless seriously injured; otherwise, it would not remain where it was. Now I decided to climb the hill from another side, try out another prospect. We were going round the base of the hill when I saw the fore part of the animal about thirty feet above where we were standing. I at once discharged a barrel at the region of its heart. Again shooting from straight below I misjudged the angle. It stumbled down on its forelegs; then raising itself up started turning to face us. I then attempted the brain shot, unfortunately, again from a sharply low angle. It collapsed as if poleaxed. After a few moments the hilltop presented a bizarre spectacle: the four legs of the tusker weaving in the air, the animal presumably lying on its back like a turtle, trying to pick itself up. I should have realized that my attempted brain shot had failed to reach its mark, but had passed close enough to knock it out completely for the time being. An animal shot through the brain does not move its limbs. I had not yet learnt how difficult a brain shot was from below at that sharp angle. Whatever Mangra thought, I had not yet emerged from my novitiate. So mistakenly confident was I of my shot that I reached the top of the hill in a run.

There was the animal sitting on its haunches like a dog, its head between its folded front legs less than ten paces away. Just as I shouted at my companions left behind to join me, the tusker started rising on its front legs, its head lifting to tower over me. I fired once again at the centre of the head hanging above. There was no time to calculate the niceties of the angle of the shot just then. The animal sank down and toppled over on its side. I had just one bullet left. It was a sad and gory end of a noble adversary.

10

~

Dead Giant
Garo Hills, October 1967

After Cachar I was away from the country for a few years. I returned in 1966 and in October 1967 I was once again in the forests of Northeast India with an elephant control licence in hand: this time in the Garo Hills.

In my childhood the Garo Hills were my holy grail. From my home on the east bank of the Brahmaputra in Mymensingh district (now in Bangladesh) the Garo Hills were about twenty miles away due north, a darker blue wash on a bright blue sky. The hill range ran from east to west, with a more or less flat top. Midway there was a bulge, sloping on the eastern side but with a sharper drop on the west. The locals called the bulge Chutmang, while the plainspeople of Mymensingh called it Kailash. This is the second highest peak in the Garo Hills, the first being Nokrek in the Tura hill range further west. The two ranges appeared as one from my home. My maternal uncle's village, Baghbed near Purbadhala railway station, was ten miles closer to the Garo Hills. Looking from there through a pair of ancient 20 X binoculars, which once belonged to my great-grandfather, I could see details on the southern slope of the hills, even a very large tree on the eastern slope of Chutmang. The flat top to the east of Chutmang, I realized when in the Garo Hills, was the Balphakram

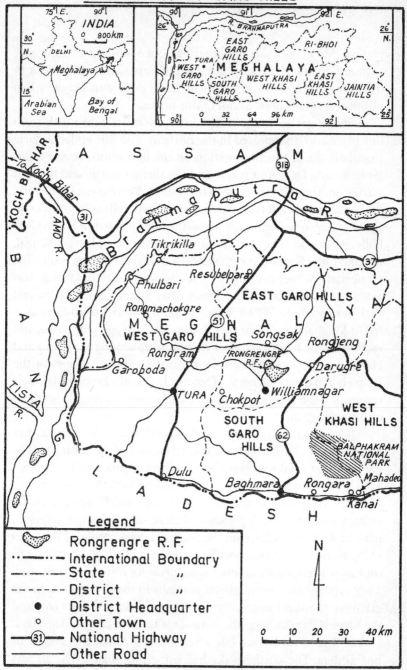

Map of Garo Hills

plateau. My maternal grandfather had once brought a young agar (*Aquilaria malaccensis*) sapling from this area and presented it to our family. It had reached considerable girth and height when political turmoil threw us out of our home.

A metre-gauge, single-track railway went straight north from our place and disappeared in the horizon. The line ended at Jaria Jhanjhail, the last railway station on the line, about twelve miles away. Susang Durgapur was four miles further north, and the hills started another four miles or so from there. Beyond the border was the weekly market of Bhangbazar in the Garo Hills at Baghmara on the river Simsong. Baghmara is now the headquarters of South Garo Hills civil district. The river, called Someswari in the plains of Bengal, is named after Someswar Pathak who founded the Susang zamindari estate in the sixteenth century. Oral tradition holds that Pathak was actually a Pathan, Samsher Pathan. After Bokainagar Fort, the last Pathan stronghold in East Bengal, about two miles from our place, fell to Mughal forces, Samsher Pathan fled, changed his name, and established by force a status of chieftainship among the Garos and Hajangs—tribal people who lived on the southern slopes of the Garo Hills in the region north of Susang. Eventually Emperor Jahangir recognized the Susang zamindari by a *farman* bearing the imprint of his palm (*panja*).

After my grandmother's death at childbirth, my grandfather married again. His second wife was a daughter of the Susang family, and Jahangir's palm imprint was their most precious possession. This is why in a way I always felt a blood tie with the Garo Hills.

From Bhangbazar, the plainspeople regularly bought exotic animals like the hoolock gibbon and racket-tailed drongo to keep as pets. In Angratoli reserve on the south-western border of the Garo Hills, my forefathers occasionally carried out kheddah operations to capture wild elephants, as much for sport as for restocking their stud. They carried out howdah shikar annually in the dry months in the extensive grassland bordering the southern fringe of the Garo and the Khasi Hills, which together now constitute the state of Meghalaya.

I still remember one late afternoon in April, the season of nor'westers. The northern sky had turned ink-black. The air was heavy and still, not a leaf stirred; the earth was holding its breath waiting for the storm to break. Panic-stricken flocks of white egrets strung out against the dark sky were desperately trying to wing their way against the strong current of air in the upper regions of the sky. Just then a brownish pillar snaked up from the ground in the far distance, clear against the dark-blue backdrop of the Garo Hills. It reached the low-hung drapery of the heavy clouds and started swaying like a headless belly dancer, slowly moving from east to west. It was a tornado performing its seasonal dance of death and deluge. People at home were quite unconcerned, and seemed familiar with the phenomenon. They stared at the northern horizon for a moment and announced casually: 'Huh! Hati sunra' ('phenomenon of the elephant trunk') and that was that. Indeed it swayed like the trunk of an elephant, thick at the top and tapering towards the ground.

All this was the very stuff of romance, and what a thrill it was to be able to come to the Garo Hills at last as an adult. The natural approach for me from the south, from Mymensingh, was now blocked by the changed political geography of the country. I had to approach it from the north, driving nearly a thousand miles—to reach a place only twenty miles from where I had grown up.

I

Road communication to the Northeast had improved dramatically by 1967. In 1961 it had taken me nine days to reach Cachar by road. Now it was only three days by road from Calcutta to Goalpara on the south bank of the Brahmaputra. The only ferry crossing was at Jugighopa Ghat, just before reaching the town of Goalpara. The Garo Hills were then in undivided Assam under the Goalpara forest division with its headquarters at Goalpara.

K.K. Barua, who was then Divisional Forest Officer of Goalpara, retired a few years ago as the Chief Conservator of Forests of Assam. Never mind your heavy rifle, he told me when I met him, always carry a shotgun with you. The year before he had been in Rangrengre, famous for its elephants and sal trees. He was walking back to camp when he met a king cobra on the road about a mile from the camp. It chased him unprovoked. He sprinted along the road, the reptile slithering after him. Mr Barua was a renowned sportsman in his days and was absolutely fit then. He ran for all he was worth up to the entrance of the camp, pursued all the way by the snake. Seeing the sahib in dire peril, his people rushed out with sticks. The thump of many running feet on the ground did the trick. The snake abandoned its pursuit and disappeared in the undergrowth.

Mr Barua also warned me that the elephants of the Garo Hills were very aggressive because most of them had gunshot wounds from muzzle-loaders earned while raiding crop in the jhum fields. All this made the Garo Hills even more irresistibly romantic to me.

I set up my first camp at the Holoidanga forest rest house, which was just a single room on short stilts next to Dibru Hills reserve, a small patch of government forest, an island of tree cover in the midst of jhum-devastated slopes. I had my first encounter with wild elephants in the Garo Hills in this reserve. I then moved to Tura, the headquarters of the district, to meet the deputy commissioner, the actual licence-granting authority under the old rules as I have noted earlier.

Tura is a charming little hill town. Though only 1500 feet high, it manages to give the impression of being a regular hill station. I remember Tura most of all as the place where I later acquired my golden cat, then a mere kitten, its eyes still cloudy. I was sitting in the shop of my good friend Nishi Kanta Deb, who dealt in motor parts, the only such shop in Tura. Additionally, Nishi Babu owned the only petrol pump as well as the only 'foreign liquor' shop in town, making him the most influential man in town.

It was bazaar day. A village Garo came up to the shop and said he had a tiger cub for sale. He was asking Rs 50 for it. He brought the tiny thing out of a wickerwork basket normally used in those parts to carry small poultry to the market. It was obviously a golden cat. It had stripes on its face, which made it a tiger for its owner. I insisted that far from being a tiger, it was only a jungle cat and pointed out that it had no stripe marks on the body. The villager stubbornly said that he was not fussy and was prepared to make do with stripes only on the face. At that moment the kitten obligingly emitted an unmistakably cattish miaow—hoarse, slightly deep throated, but nevertheless the call of a bonafide cat. I looked the villager in the eye and asked, 'What now, isn't it a cat?' Instantly a deal was struck, a ten-rupee note changed hands, and I was the proud owner of a golden-cat kitten. It grew incredibly fast, and within a year was 4'8" from tip (of nose) to tip (of tail).

The deputy commissioner at Tura, one Mr Dutta, a kindly soul, was horrified that an innocent from Calcutta was venturing into the Garo Hills all by himself and that too in search of elephants. He insisted that a local elephant hunter of repute, Hemson Marak, accompany me as my guide. Hemson's favourite hunting ground was the area around Rangrengre, mainly because it was easily accessible from Tura. I felt I did not need a guide, but one did not contradict a deputy commissioner, the lord and master of the district, over such trivia. I meekly submitted and embarked on my journey to Songsak, via Rangrengre, in Hemson's company.

The Rangrengre–Songsak–Darugre reserve forests were then connected by an uninterrupted stretch of private forests and accepted as the most elephant-infested area north of the Tura hill range. Nobody then knew much about the elephant situation on the southern slopes bordering Bangladesh. Its traditional link with what had been East Bengal (now Bangladesh) was now blocked to normal cross-border commercial activities; hence few had any interest in visiting the area and it had became *terra incognita*.

The Public Works Department bungalow at Songsak was a ramshackle two-room affair on a small hill overlooking the main road, boasting a few wooden chairs, boarded bedsteads and one cracked unvarnished table. There was no electricity or running water, of course. PWD and forest bungalows did not then believe in pampering their guests with mattresses, pillows, bed linen or mosquito nets. These were all assumed to be necessary parts of one's camping kit.

Hemson taught me how to gather information of marauding animals, a fine example of appropriate technology. One has to go to the weekly markets where people from faraway places gather and ask after their elephant problems. Garo people are usually reluctant to talk of problem elephants. It is remarkable that though the Garos practically live with elephants, they know little of the habits of the species. They were not even aware that there was such a thing as a tuskless male elephant. They have a term for the tusker, *wajal*; a term for elephant herd, *jilma*; but none for the makna (tuskless male). Yet while a Garo merely shrugged when told of a nearby tiger, elephants worried them. If they heard a tiger was roaming the roads at night, they would brandish a torch and carry on. But elephants were a different proposition and were to be avoided at all cost. The average Garo from a village believes that elephants can hear distant sounds because of their large ears; so it is dangerous to speak against them even when miles away from the forests. So respectful are they of elephants that they will never directly refer to an elephant (called *mangma*) by name, but indirectly as *dalgappa* (big fellow) or *achhu* (old thing), just as it is not considered respectful among Indians to refer to their seniors by name. If the elephant overheard your whispered words far away in the forest it could come out at night and teach you a lesson—which usually consisted of breaking down houses and destroying crops. Only when the situation became intolerable would the Garo speak accusingly of a particular elephant. You would come to know of an elephant killing people after the score had reached the tally of a dozen or more humans.

These village bazaars are not only places for sale and purchase of things, but places to socialize—a sort of weekly club meeting, where people living in very remote areas exchange gossip. I have known a Garo who would march eight miles up and down ostensibly to buy some kerosene or salt, and then spend the whole day gambling happily with bidis as stakes. Then after a night's rest at home, he would go on to keep his tryst with another bazaar date about twenty miles or so away on some other day of the week. In thickly populated Cachar, the situation had been very different. One did not have to seek out information; information sought one out.

With Hemson I went to the weekly bazaar at Songsak at the foot of the hill below my bungalow. There we gathered that a tusker was devastating the crop fields around a village named Khera, somewhere between Rangrengre and Songsak and about six miles north of the road. Obviously the situation at Khera had become critical for the local people; otherwise we would never have heard of their troubles. The next day we reached Khera, with a local hunter from a village close to Khera—carrying an unlicensed muzzle-loader—acting as our guide.

We reached the village at about ten in the morning. The village of Khera was on the ridge of a low hill. Women were busy with their daily chores. They were surprised to see us and when asked about the elephant, began screaming shrilly together. The tusker was close to the village just then, about a hundred feet down the hill. It was a revelation to me how brazen these animals could become when raiding crops without meeting any effective resistance. The animal had ignored the usual sounds and smells of the village, but perhaps smelling something different now, with our intruding presence, it started dashing down the hill. After a while I could see it climbing up another hill across a gorge. Shooting across the gorge, separated from the elephant by at least 400 to 500 yards, I settled my score. It was a fit case for control shooting: persistent damage by a lone male elephant. I paid Rs 50 to the local village chief to meet the funeral expenses of the dead tusker, which was to be arranged the next day.

The next day I accompanied the local forest staff for a survey of the carcass of the animal, as was mandatory under the rules. I found the whole village completely drunk: men, women and children, mourning the demise of the tusker. They carried the jelly-like nerve tissue that elephants have in the tusks and buried it ceremonially in the sandy bed of a stream, planting a beautifully wrought totem of bamboo and cane to mark the burial spot. This was the prescribed ritual to propitiate the departed soul of the dalgappa, so that its spirit would not haunt the village.

II

THE TERROR OF RANGMACHAKGRE

Back from Songsak, I met the deputy commissioner just to assure him that the innocent abroad in the wilds of the Garo Hills was safe and back. His immediate reaction was, 'Thank God! You are back safe.' But his official compulsions overriding his personal concerns, he nevertheless asked me to go immediately to Rangmachakgre, a village about eight miles short of Fulbari on the Tura–Fulbari hill road, about fifty miles from Tura. In that village, over the previous five nights at least one person had been killed each night by a tusker.

On the way from Tura about two miles before Rangmachakgre village, there was a PWD bungalow. I reached it late in the afternoon. As far as I can remember it was a couple of days after the full-moon night of Laxmi Puja. It was a small wooden bungalow on a low hillock on the left of the road from Tura. I asked my cook, who had come with me all the way from Calcutta, to light the kitchen fire and arrange for some tea. Meanwhile, I thought I should visit the village two miles farther down the road and inform them that official succour had arrived and that they should let me know when the tusker appeared. I planned nocturnal vigil at the village the next evening.

I drove down to the village. There was a low spur of hill, about twenty to thirty feet high, on the left of the road beyond which lay the village. The hill had been completely jhumed out and was covered with grass. Rangmachakgre itself was on almost flat land on the north-western edge of the Garo Hills, practically in the flood plains of the Brahmaputra. West of the spur of hill were parallel ranges of small hillocks jhumed out by the villagers and between the hills were fields of paddy waiting to be harvested. North of that was the village, surrounded by paddy fields. Though a Garo village, it had been influenced by the plains' practice of wet paddy cultivation. Many of the villagers could speak and understand the dialects of Mymensingh, Rangpur (now in Bangladesh) and the Assamese of lower Assam. With terror in their eyes, they said that every evening the elephant appeared and stayed in the crop fields the whole night.

Returning to the bungalow, I had just flung a tired shirt over the back of a chair, hollering across the courtyard to the kitchen for tea and buckets of warm water when a group of panting villagers from Rangmachakgre appeared in the front porch. They had run the entire two miles from their village to inform me that the tusker was already in their village. I took a few minutes to snatch the shirt from the chair, my rifle from the corner, and three electric torches from the table. It took somewhat longer to prise my driver Ahmed away from the proximity of the hot tea that he had just coaxed out of the cook. We rushed down the steps to the jeep parked in the shed below. A few more minutes and we were on our way to the scene of the nightly carnage. It was already evening. All daylight had oozed out of the sky. There was just the hint of a moon struggling behind the hills in the eastern horizon.

We stopped the jeep, as directed, on the main road to Fulbari, next to the longish strip of grassy knoll, which lay along the left shoulder of the road. The huts of the village, dimly starlit, were looming in the distance.

Directed by our local guides we crossed the hilly spur at the lower, northern end. There was a foot track running along the opposite side of the spur facing the paddy fields and the jhum-bared low hills beyond. The paddy fields were dotted with little covered machans meant primarily for watchers to protect the crop from deer and pigs. These were not being manned now after the tusker had picked three watchers from them on three successive nights. After going a short distance along the track I enquired of my guides where the tusker was. They replied in hushed tones that it was just in front of us. I was still confused. I wanted to know its exact location, if it was to the right or the left of the big cottage before us. They replied that the object before us was no cottage but the tusker itself.

I stood there for a moment as if shell-shocked. After a while the presumed cottage started swaying gently; one could clearly hear the grains of paddy being ripped off their stalks by the tip of the trunk and slowly and meditatively munched upon. Elephants do not usually uproot mature paddy plants but get only the grains by pulling along the stalk, the tip of the trunk clasping the stalk like a fist. Unmistakably, it was a very large elephant facing us, probably a tusker judging from the two yellowish patches, one on each side of the base of the trunk, gleaming through the gathering ground haze. The bridle path was about five feet above the paddy fields, which gave us a false sense of security. The elephant was facing me at an angle, but that did not lessen the visual impact of its size, standing as it was less than fifty yards away, looking like an average-sized hillock.

I stepped a few yards to my right to get the correct 90° angle.
The lesson had been well learnt at Katakhal. The excitement and
the slight gradient I had to negotiate while carrying the heavy rifle
made me pant a bit. Even in the evening chill beads of perspiration
had formed on my brow. I was determined not to allow myself to
be hurried. The tusker did not seem to be in any hurry; why then
should I be? I opened the breach of the rifle and rechecked the two
rounds in it following my usual practice of changing the chambers.
When my breathing had returned to normal, I shouldered the rifle
and pointing it to the form ahead which looked more like congealed
mist than anything solid, asked Ahmed to focus the five-cell electric
torch over my right shoulder along the barrels of the rifle. I no
longer believed in torches clamped to the barrel of a firearm. The
jerk of the recoil of a heavy rifle often makes the electrical connections
in the clamped torch come loose, plunging one without warning
into total darkness, as had once happened to me at a critical moment.

I pressed the butt of the rifle to my cheek expecting to see the
animal over the foresight of the rifle. A thick grey mist was all that
I saw through my glasses. The elephant went still and stopped feeding
on the paddy, apparently perplexed, trying to figure out what was
going on. I realized only too well what had happened—my glasses
had fogged over. In a quiet whisper I asked Ahmed to switch off the
light. Taking out my pocket handkerchief I wiped the glasses and
my face clear of all dampness. I asked my local guides to pick up the
other two electric torches and switch them on as well when I next
asked Ahmed to switch on his.

All the three torches blazed out together on the animal. There
was no doubt now that this was a tusker, though the body and the
tusks under a heavy coat of the local reddish earth looked almost
indistinguishable, except two blobs of lighter brown on two sides
of the trunk. The tusker then started approaching us in steady,
measured steps. It was by no means a determined charge at a located
quarry, but more an investigative enquiry into an unusual source
of disturbance. Suddenly our elevation, about five feet off the
ground, did not seem all that reassuring. Our two village guides
were unflinching. Although they knew what they were facing from

their experience of the last few nights while we knew our subject only from hearsay, their torches did not waver. I managed to let off both the barrels, one after the other; the first one when the tusker was coming straight towards us, the original distance of fifty yards already halved, and the second when it was veering away after the first hit. Before I could reload, it disappeared in the tall post-monsoon grass covering the hill beyond. I saw with dismay its massive stern melting away in the tall grass. Moments later I heard something crashing down in the grass. What a relief! The animal was down then! Seconds later the crashing sound was repeated. This was worrying! Once down after receiving a body shot an elephant does not usually rise again. This is not true of attempted brain shots. After a short while there was a crashing sound for the third time; and after that, finally, silence. Our two guides shouted at the village that the elephant had been hit. There was no response. Nobody came forward to meet our party. They knew the devil only too well, and now, to cap it all, it was injured as well. They had, apparently, decided to await developments behind closed doors. There was not a trace of light anywhere in the whole village, wrapped now in sepulchral silence, not even a whimper from a village cur. I told my guides that nothing more could be done that night, and that I would be back the first thing in the morning.

A depressed and a pensive party returned to the Rongmachak bungalow. Despite a gruelling day, which had involved personally driving the jeep over more than a hundred miles of rutted, twisting and bumpy hill road followed by a tense evening, I slept only fitfully that night. The vision of the great tusker haunted me. Was it going to be the familiar story of the 'big 'un that got away'?

By six in the morning it was no longer possible to linger in bed. Usually I prefer not to enter the dew-sodden forest or any kind of cover earlier than eight or so in the morning, just to avoid getting my clothes, rifle and spare cartridges and everything soaked in dew. But this was not an ordinary morning.

When at last we set off in the jeep, the sun, however, was already up. Within a few minutes we were in the paddy field, the scene of the encounter the night before. Villagers were gathered round the footprints of the elephant in the paddy field. The footprints told their own story; the exact point where the elephant had entered the crop fields and the point of its exit. Beyond the first low hillock were row after row of low, jhumed, and grass-covered hills extending to the western horizon. The phrase 'grass-covered' in the context needs explanation. This was not the well-mannered, genteel green of neatly mowed lawns and closely cropped meadows. This was elephant grass, ten to twelve feet high, which comes up in fallow jhum land during the rains and can shelter even the tallest of the tall elephants. Along the border of the paddy fields in the apron wall of tall grass were tunnel-like openings, obviously game paths. The elephant's tracks had gone back to one of the larger tunnels, its rounded top closing about four feet above the ground. A large animal passing through tall grass tramples it down underfoot to make its passage. Immediately, the grass on the two sides of the passage bends down to close the opening overhead, creating a deceptively small tunnel. The tracks of the animal disappeared into one of the larger tunnels. Who knew how far I would have to follow these tracks! I knew nothing of these uncharted private forests or how far these forests stretched. I had not forgotten the experience of Katakhal. From past experience, it was essential to have local guides; otherwise I could soon be lost in the sea of grass. But not one villager, not even my companions of the previous night, offered to accompany me. We were total strangers to them, but they knew well enough their nocturnal visitor. When I asked for two volunteers to accompany me as my guide, a double row of blank faces with blank eyes faced me.

But something had to be done. I had to try to do my best on my own with such help as Ahmed could render. How I missed Mangra! Ahmed was not my employee, but a companion by choice when I went to forests. Like me he could not resist the call of the wild.

I had to devise a tactical approach to the problem. Once I was in the grass the first problem would be visibility. Fortunately a few soot-blackened, half-burnt trees were standing in the jhumed hill burnt the previous year. The idea was that Ahmed would climb one of the trees close to the selected tunnel. In that grassy terrain an elevation should give him a wide, bird's-eye view, an advantage that is denied one in thickly wooded, or bamboo cover. As long he was up in that tree, he was safe; as long as he shouted warnings at me from his vantage point, I was also reasonably safe. I would call out to him when I reached the next tactically placed tree, and he would climb down from his perch and join me on the ground. Then the whole process would be repeated. He was to be my one-man watch-and-ward team against the five-ton terror lurking in the tall grass.

I bent down to enter the tunnel and crouched forward—there was not enough headroom to stand up straight. It was obviously a well-used track, matted with flattened stalks of elephant grass. Soon I met a narrow and dry watercourse cutting across the track. Two jungle crows fluttered up cawing from the dry streambed from a point just around the corner. My heart missed a few beats. I crept forward to the bed of the watercourse, glad of an opportunity to stand up straight. And there on my right was the dead giant of the forest: the largest elephant I have ever seen. The circumference of the front foot was 5'3", that is, when standing it must have been *at least* 10'6" tall, if not a few inches taller, and massively built, with very thick but short tusks. I have seen an elephant 11'3" high; but that was a mean-looking, lanky, leggy makna. Before me was a tusker of the pure *kumeriah bandh*, massive like a mountain peak, vast as a monsoon cloud, a mount fit for a king. It is a pity it had decided to adopt the ways of Nalagiri, the killer tusker in the Jataka stories.

11

Red Eye
Garo Hills, October 1968

The Baghmara–Rangara hill road had then been completed up to Kanai, which was a few miles short of Mahadeo on the southern fringe of what is now Balphakram National Park in the South Garo Hills district. Kanai was about ten or twelve miles farther east of Rangara. The road beyond Kanai, another five miles east, to the village Mahadeo on the Mahadeo river near the border of the West Khasi Hills district, was still being carved out of the hills.

We started from Calcutta at the end of September. Our team for the Garo Hills consisted this time of Arya Bagchi, a friend and relation we called Aju Babu, who was in the transport business; Dr Shiben Lahiri, who loved being in the forests, and I—besides the usual retinue of Ahmed the driver, and a cook.

A drive to the Northeast was seldom uneventful. This one was no exception. Just before Kulik, at around eight in the evening we saw a lorry at a standstill with a large box on the road behind it. Seasoned transport man that Aju Babu was, he instantly understood the situation. Someone was pilfering cargo from the lorry, he exclaimed, apparently a common problem in his line of business.

I was driving. I pulled up behind the lorry, got off the jeep and started shouting at no one in particular, 'Bring out my *gun*!' The fact that all my weapons were packed in boxes at the back of the jeep did not cramp my style. I was dressed in khaki trousers and bush shirt and was sporting a black beret to protect my hair from dust through our long journey. The impact of my dramatic appearance from a jeep dressed in what must have appeared to be police uniform had its desired effect. The lorry raced off, and in the penumbral shadow of the headlight of our jeep we saw three figures running pell-mell through the long paddy stalks in the fields by the roadside. Now, in a matter of seconds, there was only the box on the road before us.

We approached closer to examine it. Fortunately it was in pre-RDX days. Opening it, we found it was a large carton of 10,000 packets of Virginia No.10 cigarettes. What were we to do with it? The legally correct thing would have been to deposit it at the nearest police station. The sane voice of Aju Babu, the old warhorse, warned us against such foolishness. We, as material witnesses, would then be summoned to the police court at Balurghat repeatedly over the years—such police cases drag on and on. Non-appearance was a serious offence. What then? Selling them in a town on our way would be immoral. We could not possibly make a profit out of stolen property. Should we smoke them? But Virginia No. 10 was not our brand. Finally we decided to carry that huge carton all the way to the Garo Hills and distribute them among the deserving poor. The plan worked like magic. We were monarchs on that occasion of all we surveyed in the Garo Hills. Instead of proffering a fiver or something smaller for a paltry service rendered, we grandly gave away twenty packets of Virginia No. 10, marking new heights of munificence in the area.

We spent the night at the Kulik forest inspection bungalow. The raucous cries of thousands of open-billed storks woke us up the next day. The bungalow was obviously in the middle of a large rookery of these storks. When I came back to Calcutta I made a strident call for declaring the area a sanctuary, which was done, a mere ten years later. Later, on the insistence of a forest minister who also happened to be the tourism minister at the time, there were some attempts at beautifying the place to attract tourists; it included cleaning and deepening the adjoining water body for boating. The shallow *jheel* used to have lots of emergent water lily leaves and other vegetation until the government noticed its potential. Once cleaner and deeper, the jheel drove away the water birds including the numerous dainty Pheasant-tailed Jaçana which had made it their home. The effect, as far as we could see, was achieved without attracting any tourists.

After the usual stretch of strenuous driving we reached Rangara and its tiny two-roomed PWD bungalow overlooking the Rangara river. The plains of what was then East Pakistan were visible from the bungalow.

There were persistent reports in the local bazaar of a particularly troublesome tusker a few miles farther east at Kanai, where the PWD had a large camp for its workforce, which was building the road to Mahadeo. Mahadeo is on the southern limit of what was then only a distant dream, the Balphakram National Park. This camp was badly affected by a 'single tusker', who in this case possessed only the left tusk. Normally this would not have attracted the attention of the authorities because of the remoteness and inaccessibility of the place, which was about twenty miles or more from the Phanda forest Beat Office and 36 miles or so from the Baghmara forest range office. The entire hilly stretch had to be covered on foot to reach a bus to travel another 60 miles in one of the very infrequent buses to Tura. Only then could a complaint be lodged at the office of the deputy commissioner and the Divisional Forest Officer. But since it was the main PWD work camp at that time, the problem had received some attention from the authorities at Tura. By the time I arrived, the animal had been declared a rogue three times over already, each time for a period of six months. The period of its last notification had expired about a month before, I learnt later. After the expiry of each period there was a lull, till further complaints poured in.

I went to Kanai to gather first-hand information about the notorious 'single' tusker, intending just a preliminary survey with the objective of verifying the reports current in the weekly bazaar at Rangara. Aju Babu came along with me to savour the feel of the forest. He was carrying with him his ancestral .355 Mannlicher with no object higher than picking up something chance-offered for the pot. When I reached the place on the 1ˢᵗ of October, the camp dwellers were vociferous. The animal was a perfect pest. One fellow was walking with a bundle of salt hanging from a stick carried on his shoulder. It was snatched from behind. The authorities had supplied a muzzle-loader to the camp for self-protection. This did not help. One morning the headman, carrying the gun as his status symbol, had retired behind a bush for you know what. He was nearly pushed from behind while squatting at the job. Some of the reed huts were in a broken-down state. There were footprints on the ground everywhere and their size indicated that all of them belonged to a single animal, a ten-footer. Daylight brought no respite to the suffering people. The tusker could appear any time, anywhere, and take from the workers' huts whatever it fancied. The PWD workers said that it had not appeared the previous day. The tusker must, they said, have gone up the hills towards Balphakram plateau or gone down the hill to the plains of Bangladesh where paddy was ripening in the fields. Yes, there had been cases of human killings but we could not get any first-hand reports as the affected PWD employees had requested transfers elsewhere. Temporary labourers employed by contractors lasted for only a short time.

If the reports were right and the tusker was always hanging around the place, it was just a matter of time before I caught up with it. As it was already rather late in the morning and there was no fresh report of its presence, I decided to return the next morning to pick up its fresh spoor, if any. We had turned the jeep around and were making for Rangara when we found a Garo couple hammering blocks of stone by the roadside into stacks of small chips, an essential road-building material. I asked the man as a matter of routine if he had any recent news of the elephant. He casually lifted his hammer and pointed it to his left saying, 'Yes, it is over there.' Indeed, a closer look revealed huge footprints on the grassy left edge of the road, going down the southern slope of the hill. When did it pass, I asked. Less than an hour ago, was his laconic reply. So when we were actually interviewing people at the sprawling camp the elephant had been eavesdropping.

Getting off the jeep I asked him if he would be kind enough to show the animal to me, that is, in short, act as my local guide. Being a border Garo, he could understand Mymensingh dialect and appreciate my fulsome compliments to him as the glory of Garo youth, heroic spirit incarnate. His wife also, unfortunately, could understand the dialect. She shot up straight as a ramrod from her squatting position and started screaming at him in pure Garo. My knowledge of the Garo language is rudimentary, just enough to get by, but from the pitch and tone of her verbal torrent I could understand that she was not fully appreciative of my request to her husband. He was *not, not, not* to accompany these strangers from town on such dangerous business. However, I had done my job with the man's masculine ego thoroughly enough. A mere woman, even in a matrilineal society, was not going to stand between him and glory, so his words seemed to say. After a few minutes of futile scuffle, he firmly shook off his spouse's restraining hands, put down his menial hammer, picked up his manly dao and was ready to lead us. He left a tearful wife behind when we started our tracking.

There was in fact no artful spooring involved. The fresh tub-size tracks were clear on the ground. I had to tag Aju Babu along when entering the forest, as I could not possibly leave him alone in the jeep that was parked very close to where the tusker had crossed the road downhill from north to south. But Aju Babu, unlike Ahmed the driver, was no expert climber of trees in an emergency.

We had not proceeded even half a mile when we heard the elephant squelching in mud. It was well past ten in the morning—time for midday rest and a cooling, refreshing mud bath. The sound was coming from about a hundred yards away. I left Aju Babu and our Garo guide with his dao behind at a spot near the track, and stealthily crept forward. I had no need for a tracker any longer. Aju Babu's .355 Mannlicher loaded with soft nose bullets was more of a valiant gesture than the promise of solid aid against a five-ton adversary. There was no need, therefore, to expose them to any unnecessary risk. Thus conscience doth make heroes of us all.

The approach to the source of the sound was slow and time-consuming. As I closed up, the sound of squelching mud became clearer with every cautious step. At one point I could even hear the flapping of the ears, but of the elephant I could see nothing. Slightly farther on there was a narrow path not more than three feet wide, branching off from the left of the track and going down. As it started descending, its sides grew into walls as in a cutting leading to a causeway across a river. I then realized that the sound was coming from somewhere below the narrow path going down. The tusker was definitely down there, somewhere very close, most probably in a mud pool, but was as yet invisible. There was a crosswind blowing from the plains below to the hills in the north. It was a relief that there was no immediate possibility of the animal scenting me, particularly because I was above it as well. I inched forward, rifle on the ready, and saw a well-like depression in the ground in front. I could see the back of the elephant's head, but only its forward part. There was a piece of rock around which the track I was following went; on my left was the other path sharply sloping down to the bottom of the mud pool. I silently skirted round the rock and stood on the edge of the well-like depression. The tusker was directly below me, its massive stern about five feet from where I was standing. This was a complete reversal of the situation at Rosekandi where the tusker had been directly above me. But the correct angle for an accurate brain shot was equally problematic. The heart shot, standing as I was directly over it, was obviously impossible. The elephant was at a slight angle to me. There was the inherent possibility of a slip up if I tried the back-of-the head shot, an ideal shot for an elephant obliging enough to lie down on its side for the man with a rifle. This is the shot used to dispatch elephants sick or wounded beyond recovery. Stracey records being compelled to take recourse to such a shot on one occasion (P.D. Stracey, *Elephant Gold*, 1963). Looking back, it was an ideal position for a rump shot with a dart gun; but I did not have a dart gun with me, nor had anybody else then in India. The technology had not developed yet. Even if the technology had been available, the situation after the shot in the mud hole would still remain a problem; what would we do after tranquillizing the animal?

In a flash I decided that it had to be a back-of-the-head shot, where the two bumps fell back on the nape at the junction of the spinal column and the cranium. I know this reads like a butcher's sketch pointing out and naming the choicest cuts, but it cannot be helped if one does not mean to leave out the nuts and bolts of the business.

I waited patiently for the tusker to change its angle, afraid all the time that it might scent me any moment at such close quarters. The elephant changed its position slightly, offering me a straighter view of its back. I fired at the predetermined point at the junction of the neck and the head, aiming, hopefully, for the brain way below the point of impact. The elephant fell down on the mud with a splash. I could hear it struggling below and realized from my experience at Rosekandi that I had missed the mark. I rushed to the main rim of the wall, hoping for a view clear enough to deliver the *coup de grace* by way of another couple of shots, this time, hopefully, better placed. Instead, I saw only glimpses of its struggling legs below the lips of the overhanging wall of the fifteen-foot-deep mud hole. I rushed back to the edge of the sloping path for a clearer view of the proceedings below from that angle when the massive head of the animal loomed up along the path, its original path of entry. In a moment the head was at the level of my boots within touchable distance, its huge left tusk pointing skyward, its piggy left eye, the only one I could see, blood red. I fired almost from the hip, as they say in Westerns. It was an instinctive shot, not consciously aimed at any point in particular. The red eye guided the instinct. The bullet hit just above the eye. It was a perfect brain shot. The tusker died standing, leaning against the other wall of the passage, its left eye still open, staring unblinkingly at its exterminator.

Seconds later, after the initial euphoria, I felt shivers running down my spine when I thought of Aju Babu and my guide standing directly in the path of the animal rushing back.

I remember the dates clearly because the next day was 2 October, Gandhi Jayanti, and I had the formidable task of prising the Beat Officer of Phanda out of his lair on a public holiday. It had started raining heavily in the morning. To be fair, the Beat Officer was perfectly willing to write his report in Phanda, not trying to brave the weather. He knew his hill roads better than I did. The semi-finished road beyond Rangara on our way back was just mud and slush. The stone chips that were being crushed had not yet been laid on the surface of the road. The wheels of the jeep, even with its four-wheel drive, had no purchase on the road surface, and kept sliding sideways, showing a definite preference for an unauthorized visit to East Pakistan down below.

We had experienced local people from Rangara to help with extracting the tusk. One of them, Shingen, later became my trusted tracker in the area. We left them at the spot with our camp pressure lamp after a cursory and *pro forma* inspection of the carcass. Shingen's party came back late in the evening bearing the prize, lit by the pressure lamp. We watched the procession from the height of the bungalow, going first to the bazaar to collect local admiration, and then up to us.

It poured with rain for the whole week without stop, confining us entirely to the bungalow. We, as well as the people of Rangara, were completely unaware of what was happening in the outside world. These were the days before the invasion of the ubiquitous transistor radio. I myself never carried one when going out to wild areas, and do not carry one even now. It was only on our way back that we came to know from stray conversation and days-old newspapers of devastating floods in Jalpaiguri. After crossing the Brahmaputra we became aware of breached roads and washed-away bridges and culverts. The situation became critical when we crossed into Jalpaiguri district. The putrid smell of rotting carcasses of cattle and animals filled the air. Army outposts were diverting the traffic in the suppurating darkness by waving red lights, guiding us through unknown tea garden roads and tracks. The red eye of the Kanai tusker, appearing now and then in unsuspected dark corners, haunted me all the way in the dank darkness.

12

A Ghostly Visitor
Garo Hills, December 1968

Every year Christmas arrives and then drifts away, leaving a memory of jollity in Park Street, and plum cakes eaten and properly washed down. But my memory of Christmas in 1968 is different and indelible.

My cousin Alo (Dr S.K. Acharya, later Director General of the Geological Survey of India) was then camping in the Garo Hills, carrying out a geological survey of the area. His work took him to remote places, and he was used to camping out for months in wild country. He had with him his wife Sumitra, and a junior scientist, Santosh, to assist him in the field. As logistical support, he had a cook, a man to do the household chores, a driver, and, of course, a vehicle. He also had an elephant control licence. A wild time at Christmas was obviously staring me in the face.

On 22 December 1968 I started for the Garo Hills, laden with rifle boxes, a metal trunk, and a thick bedroll. Those days a private airline called Jamair Airlines operated between Calcutta and Assam via North Bengal. It centrally catered to the needs of the tea gardens in North Bengal and Assam and its main business was to transport people, tea chests, and baskets of pineapples and oranges to and from the gardens. Nothing was too lowly for them as cargo. Their fleet consisted of a few Dakota aircraft of Second World War vintage. These did not aspire to an altitude higher than 6000 feet and compared to the airlines of today, the bullock cart, rather than a Boeing, was the Dakota's closer kin. It took more than two hours to reach Gauhati (as it was then called) from Kolkata by Jamair, and more than one hour to reach Dhaka. The runways used by this airline had been built for temporary use by the air force during the Second World War. There was no fuss over the maintenance of these airstrips made of strips of corrugated metal sheets with perforations. When in the course of nature, grass sprouted through the holes in the metal sheets, local villagers were employed to cut it down manually. The so-called airports were only tin sheds. They were empty the whole day but for the hour or so that the airline used them. Villagers helped to load and unload the luggage, and chase away grazing cattle with sticks when the strip became operational for take-off or landing. Everybody knew everybody, and everything was informal both in the airport shed and in the aircraft. It was considered perfectly normal for pilots and crew to carry private messages and medicines for people from one 'station' to the next.

Taking off from Dum Dum, the first halt was the private landing strip at the Telipara tea garden in Jalpaiguri in North Bengal. Some baskets and some supplies for the tea garden were offloaded. A few passengers got off and a few got on. Rearranging and removing some seats created the space for new passengers and fresh cargo including some chicken in baskets as well as goats in boxes built of slats with gaps in the side for light and air. The goats bleated lustily through the gaps in the crates, adding to the air of general bonhomie. Some additional chairs for the new human passengers were set up as is done at weddings and other improvised festival venues.

The next stop was Rupshi in what was then Goalpara district in Assam. The airstrip, in the middle of a minor sal forest, had been improvised by the air force during the Second World War. This was my stop. An airline bus carried the arriving passengers to the bus stand in the town of Dhubri on the north bank of the Brahmaputra, picking up some local passengers on the way. Then, a cycle-rickshaw to the ferry ghat. The original plan had been to cross the river in a launch and then take a bus to Phulbari Bazaar on the north-western corner of the Garo Hills; and from there, eventually, by some form of public transport to reach Tura. An uncomplicated plan, no scope for any confusion: straight by plane from Calcutta in the morning, and Tura by late afternoon the same day. Private buses apparently plied straight from the ferry ghat on the south bank to Tura via Phulbari Bazaar. At Tura my cousin Alo would be waiting for me at the Circuit House. A perfectly straightforward arrangement with little scope for deviation, or so I had imagined in my serene *naiveté*.

The first jolt came at the Dhubri ferry ghat itself. I found that since at that time of year the six-mile-wide Brahmaputra did not have enough draught for large passenger launches, the service had been suspended. However, I saw many country boats setting out from the river bank, vociferously soliciting passengers. With sail and pole, I was sure, one of these would get me across in romantic style. With four boxes for weapons, another for ammunition, one trunk for clothes and a bedroll fat with blankets, I got on to a boat. I cannot claim that I was a light traveller.

With the dry winter season, the river had spilt into many channels interspersed with sandbanks. The boat had to weave its way from channel to channel negotiating these sandbanks. But although the river was running low, the Brahmaputra, even in the dry season, is awesome in its breadth. At one point I saw the carcass of a dead animal floating down the river. A pair of shelducks, the emblem of eternal love and romance in classical Sanskrit literature, were sitting on it side by side, tearing at the bloated carcass and gobbling great chunks off it. I was a keen wing shot those days, but never did I remotely consider a shelduck for the table after that. If a shelduck, why not a vulture? At least the vulture had more meat.

Balancing on the piled-up luggage in a light, flat-bottomed boat, tilting even in the sluggish current from side to side, I reached the other bank in three hours. It was almost dark. Phulbari Bazaar, where the Garo Hills begin, was another eight or nine miles. I found the last bus had already left. Was I then fated to spend a freezing December night under the open sky on the sandy, treeless wide bank of the river? After taking my luggage off the boat with the help of the boatmen, I was just looking round to assess what was in store for me when I noticed a lorry standing on the sandy river bank loading some sacks brought by another boat from across the river. As abjectly as I could, I pleaded with the driver. At last he agreed to give me and my luggage a lift up to Phulbari Bazaar, which was his destination. Putting my luggage safely on the sacks of potato at the back and seating myself next to the driver I made it to Phulbari.

Phulbari was a familiar place. During my stays in Holoidanga, Phulbari Bazaar supplied my needs. Only the previous autumn, I had spent a night in the PWD bungalow on the Jinjiram river near Phulbari. I still cherish the vision of its ripples twinkling in the moonlight. But a winter evening is not a lively time in a small place like Phulbari. Dim oil lights were flickering in grimy and yellow lamps at a few shops. On enquiry, they told me that the last bus for Tura had left and there would be no bus before the next morning. Where then was I to spend the night? The PWD bungalow was two miles from the place. It looked like I would have to spend the night on a bench in one of these shops. I realized that I had not eaten anything after leaving Calcutta. Rupshi 'airport' did not offer such luxuries as tea, or samosas or biscuits and the Jamair Company did not believe in spoiling its passengers with food packets in flight. In the rush to catch a boat about to leave, I had had no time to look for a snack at Dhubri either. As the evening air got chillier by the minute I began feeling the cumulative effect of the enforced day-long privation. These few shops did not offer any food, only tea. There were some ancient nimkis, local exotics, no doubt brought from faraway places, reposing in glass jars alongside mouldy nankhatai, the last resort of the weary traveller. After three glasses of oversweetened tea and about half a dozen nimkis and nankhatais, I was about to request the shopowner for some sleeping space in a corner of his shop when a jeep crammed with travellers arrived from Tikrikila, a somewhat bigger place about three miles beyond Holoidanga. The jeep was bound for Tura, and they were taking a break on the way for a glass of hot tea. The boisterous crowd, laughing and singing, seemed to have already had stimulants stronger than tea to ward off the cold. The driver agreed at once to take me to Tura.

In that mood he would have happily agreed to drive me to the summit of the Everest. He even found a place for me in the front seat. That made six of us in front including the cheerful driver. My fellow passengers thought it a great lark to carry my pieces of heavy luggage on their lap. My left buttock was on the seat, the rest of my body hung outside; it was too precarious for me even to try and imagine how the driver was managing at the other end. In course of time the jeep started for Tura.

The road from Phulbari winds round the western end of the Garo Hills where the old man Brahmaputra, tired of rolling on and on from east to west for so long, takes a plunge southward, at last freed from the retaining wall on the left. The road hugged the base of the hills on its left; on the right were extensive paddy fields in the floodplains of the great river. A few miles past Garo Boda Bazaar on the way, the road starts climbing to Tura. All the way before the short climb to Tura, it is hills on the left and paddy fields on the right. At the end of December the harvesting season was nearly over but sufficient unharvested paddy still remained in the fields. All the elephants in the region, numbering about 500 or so according to the latest count (2002), gather on the edge of the hills and descend upon the paddy soon after, or even before, dark. This area, Dadanggre, is one of the places worst affected by man–elephant conflict in the Garo Hills even today.

We were a merry party on the way to Tura. The road hugging the outer edge of the hills dipped now and then in deference to the numerous unbridged streams that debouched out of the hills into the plains. Each dip registered sharply all over my back. Shortly we should be in Tura and my hours of agony would be over, or so I thought. The chap upstairs must have smiled into his long white beard hearing my thoughts. We were halfway through our journey. Our jeep had just descended into one of the dips when, trying to go up the other side, the engine started spluttering, and then hiccupped to a stop. The driver pulled on the self-starter a few times; the engine was unsympathetic. There were some vigorous attempts to crank it up manually. The engine was indifferent. Then the driver exclaimed with an air of revelation: he had forgotten to fill petrol when starting from Tikrikila. He said he was sorry, and it was bad business (*bea hol*). All agreed that it was indeed bad business (*bea*). He said he would have to go back to Phulbari next morning for fuel and bring it back in a jerrycan. There was a village, the driver said, once we climbed up the bed of the nullah, where we could rest for the night. The logic of the argument was ruthless.

We now had to push the jeep for about half a mile up the slope to reach the promised sanctuary. We all got off the jeep to supplement the missing horsepower with manpower. The jeep was standing on the bed of a dry nullah, paddy fields to our right, and the hills to our left. As soon as I got off the jeep I realized that it was not a safe place at all. It was thickly strewn with elephant dung boluses of all sizes, shapes and age. This obviously was one of the elephant highways from the hills to the paddy in the plains. The road came down steeply from behind and went up equally steeply in front; so there was nothing to choose between them. One might as well try to go forward up the steep incline—at least there was the promise of shelter in a village there. Putting our hands together we pushed the jeep up the incline and then further on to the village about a mile away.

The villagers came to receive us with shouts, sticks, spears and flaming torches. These villages on the border of Bangladesh are extremely vulnerable to transborder robbery, and villagers, therefore, are on the alert at night. Fortunately for us, once they were satisfied of our identity, they were extremely hospitable. Determining my social status from my cap and the amount of luggage I was carrying, they offered me their most *respectable* accommodation: the single-room village primary school. This appeared fitting to me, as I was then, professionally, an academic. But on entering the place, the idea looked less attractive. Half of its corrugated tin roof was missing, blown away by storms the October before, and not replaced yet as the rains were not expected soon. Split bamboo with wide gaps made up the sides of the shed, rendering windows superfluous. Obviously the bamboo was not meant to resist wind. In fact they broke into a merry whistle when a sharp wind blew through them. Their main object appeared to be to allow unimpeded passage of light, rain and air. The school was actually an extension of the communal village cowshed. Peeping into the next room I found it more to my taste. Cows were huddled cosily in one corner, all tied to their stakes. Freshly threshed stalks of paddy were piled up on the other side of the shed. The corrugated tin roof in the cowshed had no gaps, and its bamboo sides were in reasonable shape. It had less raw nature and more warm comfort and bovine camaraderie. Making my choice at once, I pulled out some dry paddy stalks from the pile, spread them on the ground thickly and placed my bedroll on it, adding an extra blanket on top and keeping all my weapons next to me. Sleep was fretful. A holy smell of cow dung and urine permeated the place. My bovine roommates, without any apparent reason, made scuttling sounds from time to time. I was later told that they were only trying to get their feet out of the way of scurrying rats, and, possibly, the slithering things that were after them.

The stubbles of sheared paddy in the fields had not yet peeped out of their winter blanket of ground mist when there was the honk of a lorry coming down from Tura. Our driver had been up with the earliest insomniac lark. He sprang forward waving an empty jerrycan and soon he was on his way to Phulbari. He returned about two hours later cadging a lift off a Tura-bound lorry. Around mid-morning, I was finally in Tura, which I should have reached in the late afternoon of the preceding day.

After a whole day at Tura where Alo, Sumitra and I collected provisions and ran errands, we were to head out to Alo's campsite, a place Sumitra was not on the friendliest terms with. Alo had put up his tents at a place called Siju Arteka on a prominence overlooking the river Simsong. The government supply tent, expertly put up under Alo's supervision, was apparently adequate and comfortable enough. It even had a canvas enclosure for a bathroom; but water was a problem, which had to be brought up in buckets from the river flowing quite a distance below. However, after a good deal of cajoling and financial inducement, some villagers had been persuaded to bring up a few buckets of water from the river. When Sumitra, Alo's wife, entered the enclosed area for a well-deserved wash, she found a barking deer in the bath enclosure. Startled, it wanted to escape but in the process upset the buckets of water and spilt the precious fluid brought up from below with so much trouble. The villagers gathered round, caught the animal and took it away. That night Alo's party could hear a great feast going on in the village some distance away while there was not much water in the camp except some drinking water in flasks.

There was no town of any note along the sixty miles from Tura to Baghmara bypassing Dalu. Nothing was available by way of provisions except ash gourds, masses of them, and coarse rice and salt. This, Sumitra acidly said when we rendezvoused in Tura, was where we were headed.

We had to procure everything we required at Tura, including tea, powdered milk, sugar, biscuits, vegetables and atta. Rangara was another 24 miles beyond Baghmara after crossing the river Simsong. Baghmara was the only place where petrol was available in the entire 84-mile stretch from Tura to Rangara. Petrol, sold in tins at Baghmara, was laced liberally with water, mud, kerosene and burnt engine oil. To play safe, we filled the jeep with petrol at Tura and piled in jerrycans of fuel.

The road up to Dalu, thirty miles straight down south, was reasonable; but when it took a sharp bend to the east and went along the southern border of the Garo Hills to Baghmara on the river Simsong along the border with Mymensingh, it became abominable. The road to Rangara, another 24 miles, had been newly constructed and had all the hazards of a newly constructed hill road. The river Simsong had no bridge yet. People and vehicles had to be ferried across, and the ferryman was rarely sober after sundown. Across the river started the Baghmara reserve. There was no public transport system east of the Simsong. Beyond the river, a forest Beat Officer was the sole representative of the government at Phanda, about 12 miles east of Baghmara. Rangara, another 12 miles from Phanda, was the seat of an assistant engineer of the PWD, unquestionably the highest-ranking and the most powerful officer of the government in the area, presiding over the only significant public activity in the region: the construction of a new road through the hills from west to east parallel to the old road in the plains along the border with the districts of Mymensingh and Sylhet in East Pakistan. This road had been abandoned because of the constant threat of rifle and mortar fire from both sides of the border as a form of light-hearted political fun and games, reminiscent of cross-border firings along the Line of Control in Kashmir in recent times.

I had had nothing to eat after leaving Tura in our rush to catch the ferryman at Baghmara still sober. I was feeling in my bones the cold north wind blowing down the Simsong. After crossing the river, about five miles from Baghmara, we came across taungya fields on the left of the road. I asked Alo, who was at the wheel, to pull up by the roadside. I wanted to show him where I had dealt with a marauding tusker the October before in a depression by the roadside. A half-hearted moon was playing bo-peep among wisps of cloud. As I was pointing out the spot to Alo from the jeep, I saw a largish mound in the same depression. Immensely surprised, I peered closely at the mound when it seemed to move slightly. I asked Alo to switch off the engine. There was a distinct sound of pulling at vegetation coming from the animated mound which certainly deserved a closer scrutiny.

Getting off the jeep Alo and I approached the edge of the road for a closer view. All my powerful electric torches were safely packed away in a box. Alo had only a two-cell torch in the glove compartment of the dashboard, which he brought out. His rifle was safe in its dust cover; mine were secure in their jeep-proof hard boxes. The cartridges of Alo's rifle were in his grip buried somewhere in the luggage and bags of provision piled high at the back of the jeep. We approached the edge of the depression. We could now clearly perceive the mobility of the mound. Its vague outline in the dim moonlight from a distance of about thirty yards had a definite elephantine suggestion. All this was unbelievable: another lone tusker after three months at exactly the same spot! My teeth were chattering in the bitter cold. This was our first venture in the forest together. How would Alo know the reason for my audibly chattering teeth: an empty stomach giving an extra bite to the cold; it was hardly the right image for a man aspiring to be the guide and mentor to a young man going to meet his first elephant.

The ghostly shape suggesting an elephant was uncanny. Was there something then in the Garo belief in an elephant's spirit haunting the place of its death? I remembered we had not performed the prescribed rites after the death of the elephant the October before. After all, anything was possible in the Garo Hills, its southern slope then wrapped in the primeval unknown.

I asked Alo to switch on his torch. In the wilting yellow pool of torchlight there was indeed a tusker standing in the hollow about thirty yards away. It stopped feeding as we were looking on, something like Dorothy Wordsworth's cow. We withdrew discreetly, deciding to check on this weird phenomenon the next day. Then we were on our way to Rangara and a hot meal. Just round the bend was a small shed on short stilts, a cover against the elements for vigilantes to ward off small game damaging the crop. There was a blazing fire with a group of people squatting around it. I asked them warily about the tusker wondering if they would confirm it was a troubled spirit. But they were vociferous about its this-worldliness. It was a nightly visitor and a scourge of their crops. It appeared around seven every evening and lingered over their crop the whole night. Again one could not help wondering at the wondrous place that was the Garo Hills, where even elephants were so punctual! We promised to come back the next evening at the appointed hour.

The next evening we arrived from Rangara, twelve miles away, a few minutes earlier than the appointed hour with the tusker. Sumitra and Santosh also came with us. The villagers, already assembled, directed us to a spot overlooking the U-shaped depression lying parallel to us, the right, open end sloping down to Phanda valley; on our left was the closed side of the U. There we discovered a small machan on stilts, about four feet off the ground, very close to our observation post the previous evening. I had assumed that our visitor would appear at the same spot, in the hollow from the open side of the U or across the low, hilly upper arm of the reclining U. Just then came sounds of breaking of twigs on the low hill on the other side of the hollow, the farther arm of the U. I hastily persuaded Sumitra and Santosh to climb up to the flimsy machan; its height, about four feet off the ground, however negligible against the might of the adversary, nevertheless gave a false sense of security. But the tusker, true to the reputation of the species, was unpredictable. We were expecting it to climb down to the depression across the low rim of the U on the far side; instead it decided to come along the rim of U, the sound of snapping twigs and bamboo clearly signalling its progress in the still night. It moved steadily all along the rim of the U towards the machan. Alo with his rifle (a .404), and I with a powerful torch in my right hand and my own rifle in the left, were standing immediately in front of the machan, expecting the elephant to descend to the depression in the ground between the two arms of the U; but it continued straight towards the machan along the lower arm of the U.

Suddenly I became very uncomfortable about the presence of Sumitra and Santosh hunched in the machan just next to us, giving us a running commentary in hoarse whispers on the step-by-step movement of the tusker, cautioning us repeatedly and quite unnecessarily. The false sense of security the machan had instilled in them made them worry more about the safety of the people on the ground than about themselves. I waited till the last moment for the elephant to change direction and go down into the depression between the arms of the U, but the cussed animal did not; it stuck to its course. I had decided not to allow it to come closer than a bush about ten yards away. When it crossed the *laxman rekha*, I asked Alo to get ready. Elephants in jhum fields are so accustomed to human presence that they tend to ignore it altogether. One can even smoke sitting in a machan. The animal will at best lift its trunk in the direction of the machan to indicate that it has registered the irritant—just a courteous gesture, nothing more.

I switched on my five-cell torch. Its circle of light showed a tusker about fifteen paces away, trying to gauge the situation. The head was slightly, ever so slightly, turned to its left. Acutely aware from past experience of the crucial importance of angle in a head shot, I asked Alo to aim at a spot between the central bump and its right eye. I had no time to explain why. Years of past coaching about the vital spot being the centre of the head (i.e., the bump) overruled the whispered voice of experience. Alo took a shot at the centre of the bump. The elephant tumbled down the rim to the hollow on its left and lay still. I held the torch firmly focused on the shape lying on its side. There were congratulatory shouts. We were watching the scene in the depression from the raised rim of the U. Just then I saw the ear on the side of the head free of the ground flap once. The lessons of Rosekandi flashed through my mind. I almost shouted at Alo to take care. In a few minutes the momentarily stunned elephant was climbing up the other arm of the U on the far side. A couple of well-directed shots concluded the affair.

The elephant, crucially, did not get away, and we did not leave an injured elephant close to human settlements, or turn a marauder into a killer. The expedition was not well begun but it eventually ended well, at least for us. The tusker had to pay the wages of the sin of going doggedly after forbidden paddy.

13

The Makna of Mahadeo
Garo Hills, December 1969, April 1970

There was news of a makna at Mahadeo, a few miles farther east of Kanai. This had been so far an unknown and unknowable territory for us: the road from Kanai to Mahadeo had only just been completed. When I reached Mahadeo, I encountered the makna's footprints everywhere. There was no one in Mahadeo Bazaar, close to the border of East Pakistan, who had not had an encounter with the animal. For the last two or three years it had reigned unchallenged, tyrannizing people. It had killed many men and destroyed many houses. I could see the remains of some houses it had broken. It had never been declared a 'rogue' officially, mainly because there was no one to make a written complaint against it to the authorities. There was no one in Mahadeo who could write an application. The nearest police outpost, competent to register a First Information Report, was forty footslogging miles away along the hill road at Baghmara, and the Phanda forest Beat Office was also nearly thirty miles away; but these pillars of local administration could not possibly act on the matter: it was too big for them.

I had set up camp at the Rangara PWD bungalow in December 1969 with my wife Sheila, Alo, Sumitra and Santosh. We received news that the makna was regularly invading fields of tapioca.

Tapioca, apparently, was more irresistible to the local elephants than ripening paddy. Alo and I decided to keep a night vigil in the fields, sitting on a jhum-burnt log outside a hut destroyed by the elephant. A bitter north wind was blowing down the valley of the Mahadeo river. Santosh, to protect himself against this wind, entered the abandoned hut but rushed out after a few minutes with agonized cries. In the light of our electric torch, we found his body covered with ticks, as also his clothes and his woollen cap. These were thick with ticks and bugs of every size and shape. Both Alo and I had encountered ticks in the forest before this. Ticks are much more of a nuisance than ordinary inept leeches that gorge themselves too much to hold on to their perches. One of Alo's eyes had once swelled and turned a terrifying red. We washed the eye with salt water and applied eye drops but there was hardly any relief. Eventually, after a close examination of the eye with a torch, I found that there was a little tick, cosily settled in the eyelashes of the upper eyelid. Leeches in the rains, ticks in the dry season and elephants in all seasons, these together made up the Garo Hills.

That freezing night we asked the owner of the adang if he could give us a glass of hot tea. No problem, he said. Lights could be seen in several similar low crop-protection machans dotting the slopes. Cupping his hand around his mouth and turning in the direction of a nearby machan, our host yelled, 'Winston Churchill!' We gasped at him baffled. But an offer of red tea was shouted back by Mr Churchill, whose full name, it turned out, was Winston Churchill Sangma. The missionaries had run out of names as usual; I have since met one 4 July Marak, obviously a man in the clutches of a zealous American missionary at birth.

That year, however, I did not get the makna. I visited the area repeatedly in search of it but never had a face-to-face encounter, though its footprints and the evidence of its misdeeds were everywhere, along with reports of fresh human casualties.

In April 1970, I was determined to make good use of a short vacation and see the matter to its bitter end. Mahadeo, meanwhile, had improved its amenities somewhat. The PWD had constructed a two-room structure for the public health department, but the health department had not moved in yet.

The place was a structure without a stick of furniture, not even a bucket in the bathroom. There was, of course, no question of running water. Two sanitary lavatories, one attached to each room, were the only frills the construction boasted. The doors and windows, made from unseasoned local timber, had warped already. None of them closed properly. The gap between the door leaves and the doorjamb was at least an inch wide. The only way to keep prying jackals out from one's room was to tie up the doors and windows at night from inside the room by passing a length of coir string through the pull-rings attached to the doors and windows. The chowkidar asked us to be careful, as leopards had recently taken some of his beloved goats, and two dogs from the veranda of the bungalow. We spread our bedrolls on the bare floor; the inch-wide gap between the door leaves and the jamb allowed free access to the creepy crawly things that abounded just after the rains. Anxious enquiries revealed that no phenyl was locally available to sprinkle on the veranda and on the ground around the house. We fervently hoped that mosquito nets firmly tucked into our bedrolls would be sufficient to discourage unwelcome nocturnal visitors.

The next morning I received the news that quite a bit upstream of the river, the makna raided a village called Ramphachiring every night. Shingen, Alo and I reached the village following a steep bridle-path from Mahadeo. There were signs of the makna's misdeeds everywhere. Its familiar footprints were ubiquitous, besides broken houses and devastated crop fields.

The whole day we tracked the makna all over the hill slopes but never caught up with it. After dusk, we came back to an adang in Ramphachiring, the site of its nightly depredations. We decided that we would not go back to Mahadeo that evening. The elephant was certainly hanging around close to the adang. We would wait in one of the hutments of the adang for its visit at night. For seasons on end, we had been after it, but this time we were determined to force a final showdown. The only problem was that we had no food or water with us, and the water flask that Shingen was carrying for us was empty. We would have to do with the water from the chiring, adding a couple of water-purifying tablets to it. We knew we would be invited to spend the night in one of the temporary crop-protection huts and our Garo hosts would receive us warmly.

After a whole day of tracking, carrying a heavy rifle up and down slopes and ravines, all we felt was sheer exhaustion. We made it to a family hut at the centre of the jhum slopes, and, as expected, received a warm welcome. This was not a permanent village hut, only a makeshift construction for guarding cultivation throughout the season. Our Garo hosts accommodated us in the front porch of their long family quarter. Shingen disappeared behind the screen to the interior of the cottage and returned in a while to inform us that they had no rice, but they could offer us some boiled thabalchu. We agreed heartily to the suggestion, and made ourselves comfortable on the matted floor, lying flat on our back. We had dozed off when Shingen came back, woke us up and offered us on separate enamelled plates boiled thabalchu from the fields, green chilli and salt. After asking our host to give us a call if the makna appeared at night, we gobbled up the fare offered to the last morsel and sank into oblivion using our hunter boots as pillows.

There was no alarm at night. We woke early in the morning, and fell upon our breakfast. This was now a delicious variation on the theme of thabalchu: boiled and sliced thabalchu with salt and green chilli. Then saying goodbye to our hosts, we were off again in search of the makna.

Soon after starting our search for fresh tracks, we received infor-mation that the previous night the elephant had worked out a very satisfactory arrangement for itself: as we slept at Ramphachiring, it had shifted its night-foraging ground to the adang of the village Hatishia, on the left of the Mahadeo–Balphakram road, in the catch-ment of the river Kanai. Ramphachiring, on the right of the road, was in the catchment of the river Mahadeo. Receiving confirmation of the information from some villagers on the road, we crossed the ridge to Hatishia. We confronted the same scene of nocturnal depre-dation we had seen at Ramphachiring: devastated crop fields and hill slopes strewn with broken hutments. We picked up the fresh spoor from the village and again started tracking. We were carrying in our pockets some pieces of raw thabalchu brought from Ramphachiring.

I do not know about others, but I was certainly seeing red, were surviving on thabalchu for two days, and having tracked the animal unsuccessfully over nearly three years.

On the road from Rangara, about a mile and a half from Mahadeo, a motorable track branches off north to Balphakram plateau, originally constructed by contractors to extract and carry processed agaru (*Aquilariu malaccensis*) and unprocessed logs from the southern slopes of Balphakram plateau. Uncared for, some quick-growing trees had taken over the middle of the road, and the whole strip was overgrown with bushes and scrubs. All this is now a part of Balphakram National Park in South Garo Hills district close to the border of Bangladesh, and there is a well-maintained road to the plateau. I am speaking of times years before Balphakram had acquired the status of national park and could boast a helipad for the visit of occasional VVIPs.

The place Balphakram occupies in Garo culture is something like Mount Olympus in Greco-Roman mythology or Mount Kailash in the Hindu tradition. This is the home of the Garo animistic gods: two to three hundred square kilometres of desolate flat land devoid of human presence as well as of trees, except in some depressions or cracks in the rock which had collected soil washed away from the plateau by rain. The plateau itself is the creation of table erosion with a very thin layer of topsoil, unfit for growing any kind of crop and covered with short grass, giving the appearance of a flat meadow. That is one reason why the area is not jhumed, and it was easy to acquire the land to set up the National Park.

On the southern slope of the plateau grow agar trees. When the trunk of an agar tree receives some injury, a sort of infection grows within, stimulating the secretion of a highly aromatic resin that creates a heady scent in the surrounding pith or wood. This infected part of the tree is very valuable and is used to make perfume. Its main demand is in the Middle East. Those days the district tribal council used to lease out 'mahals' to contractors to extract agar. It was customary for a Garo to take a swipe with his dao at an agar tree met in passing. He was just helping the tree to grow its scent. The contractors, after examining the pole for signs of insect attack, would cut down the tree, and burn it on the spot to extract the resin. I have seen from the Rangara–Mahadeo road agar kilns blazing on the slopes of Balphakram. Excessive exploitation has put this species on the endangered list and its export is now prohibited. The system of agar mahal has also been discontinued, but the business continues to thrive with materials smuggled from Burma.

By 1970, the agar mahal had been suspended. The road made by the agar contractors had not been maintained for some time and was becoming a part of the surrounding forests. I am describing the road in some detail as my story builds up around this road.

It was now clear that the makna was using both sides of the road, Hatishia and Ramphachiring, as fancy took it. The whole day we combed through the cover on both sides of the road. The elephant was criss-crossing the road from one slope to the other, but remained elusive, as usual. We sheltered for the night in a rickety, open crop-protection machan at Hatishia. That night dinner was thabalchu roasted, not just boiled, with salt and green chilli. A piece of a raw onion was added to the dish making it a luxury. Early in the morning, two villagers from Ramphachiring on their way somewhere informed us that the makna had visited the village the previous night. We wearily returned to Ramphachiring and again picked up the spoor from there. By now, I had come to know its footprints a shade better than the back of my hand. There was no room for any mistake in recognizing the single pair of tracks.

In the afternoon that day from a ridge I was on, I at last spotted the back of two elephants on the Mahadeo side of the road. They were browsing in a clump of bamboo below the road. Nobody had told us at any of the villages of *two* elephants, nor had we seen footprints of *two* elephants anywhere. A close watch on the two elephants soon revealed that one of them had medium-size tusks, while the larger animal was a makna. That was obviously the makna we were looking for.

After observing the pair for some time, we realized that we could do nothing from our elevated position; we would have to go down to where they were lurking. A few cautious steps at a time, we descended the slope. Crouching low and at our wariest, we entered the large clump of bamboo. There was now the danger of our steps crunching on the dry bamboo leaves on the floor. The elephants were invisible to us. The bamboo was in a valley-like fold between two hill slopes. The elephants must have entered the patch through an open end of the valley. There was every possibility of our losing direction in that dense bamboo; we inched forward guessing at the elephants' position. Just then, I heard a bamboo stem snap quite close. Now the job was to target the sound.

As the elephants broke or pulled down bamboo, we proceeded synchronizing our movement with that sound. My readers must be familiar by now with this method of stalking, never forgetting, of course, the direction of the wind. Fortunately for us there was no crosswind in the middle of the day in that hole of a narrow valley. As we crawled forward on our haunches, the sound from the bamboo ahead became clearer every moment. Now I could hear not only the sound of the breaking of bamboo, but also of the elephants munching, and their flapping ears. They were very close, ten to fifteen yards away at most. Two sets of sound were coming from approximately the same spot, so the elephants were still together. At that moment I suddenly saw a trunk snaking up and grabbing the top of a bamboo sprouting fresh leaves, and pulling it down noisily—followed by the sound of contented munching. I could see the flapping of ears now through the network of bamboo twigs. They were perhaps about ten yards from us.

I sank down on the ground. On a signal from me, both Alo and Shingen followed suit. From that low angle, sitting flat on the ground, I could see their pillar-like legs. But what was the use seeing only their legs? After all, they were not Marlene Dietrich's. Very careful of the network of twigs of bamboo over our head, Alo and I stood up for a better view; but alas, there was only an impenetrable screen of bamboo leaves and twigs blocking our view, behind which the animals were happily munching away and cooling themselves noisily with their ears. So far, we had been able to see the tip of the trunk of one animal and their legs from the knee downward.

Time was ticking away, and the elephants showed no sign of moving. Should I go forward a bit more? But something inside me said, enough, enough, no further—no doubt the result of repeatedly reading of the 'sixth sense' in Corbett's writings. I decided to listen to my inner voice. What was an impenetrable wall of bamboo to us, was merely tall grass to the elephants and they could smash through it at will. Besides, there were two elephants, one of them of established notoriety. We squatted for an hour within ten yards of the maljuria pair, but failed to get a clear view of the animals. Around four in the afternoon the elephants started moving again, northward now, along the right side of the Mahadeo–Balphakram road. This was our chance. Balphakram was to the north of where we were. I could see a narrow ridge going down from the Balphakram road to the valley on the right. If the elephants held true to their course, they could cross the ridge. If we went up to the Balphakram road and took a stand on the ridge a few hundred yards farther north, we thought we had a chance of intercepting the animals in their northward movement. We scrambled about a hundred feet up to the road, and then ran to the starting point of the ridge on our right and followed its gentle gradient down. In the post-jhum period at the end of the monsoon, coarse grass as tall as twelve to fourteen feet covered slopes of the ridge, but the top of the ridge itself had a light tree cover with a reasonably clear view of ten or fifteen yards.

The plan was to anticipate the point where the maljuria pair might cross the ridge, and wait there in ambush for them, as a jungle cat waits in cover of its choice to pounce on an approaching prey.

Just then, we heard a twig snap. In our eagerness to locate the path of the elephants from their sound, we had taken our stand on the edge of the grassy slope, leaving the narrow wooded clearance behind us—an overeagerness that could have cost us dear. Had we stood back ten or fifteen yards from the edge of the grassy slope, we would have had a clear field of fire in front of us. Now, with the elephants so close and dry autumn leaves matting the ground, we could not change our position. Even the slightest of noise would turn away the wary creatures. We were standing in a very awkward position; I in front had my right foot forward and one step low on the slope and had frozen in that position, literally caught on the wrong foot. Meanwhile, the unhurried sound of breaking twigs and dry grass being munched was steadily closing up on us, not more than ten yards away now and inching closer every second. We had anticipated their crossing path only too well. Alo was just behind me; Shingen, I saw from the corner of my eye, had dropped down on his haunches making himself as inconspicuous as possible. One elephant was so close that I dared not shift my right foot back for a more stable stance, for fear of rustling leaves. After nearly three long years I had come in contact with our quarry and was not prepared to take any chances. I glanced back at Shingen and I still remember his flared eyes and ashen face. Yet though he was armed only with his *mongreng*, he remained as still as a rocky outcrop.

One elephant was very close, the other slightly behind. I could see the tips of the tall grass shaking as they were being pushed apart; I could hear the heavy breathing of the animal, the flapping of its ears, but could not catch a glimpse of the elephant itself. I pushed up the safety catch of my rifle: there would be only a split second for action when the animal did come in sight.

The clump of elephant grass just in front of us started shaking; its tips bent. The wind was in our favour, about the only plus point in our situation. After all these years, I still remember noting the direction of the warm valley wind.

I was staring skyward, as if looking for a bird on the branch of a tree, for I was expecting the head to break out of the grass above me. At last even the last clump of grass screening us off from the elephant started shaking. Suddenly from the corner of my eye I saw the tip of the trunk, about three feet from the ground, coming out of the grass and coiling round the clump of grass. It was three feet or so from where we were standing. At last the critical moment had come, but the elephant was not visible somewhere overhead as expected, but nearly at the ground level. Even the rawest of raw greenhorns would know that the tip of an elephant's trunk was no place to hit an elephant, however close or distant it might be from you.

Just then the trunk stopped midway, left the grass, froze for a moment, and then started snaking towards us up to touching distance: the head and the body of the animal were still invisible behind the screen of grass. Then the tip of the trunk flared out, moved slightly to one side and pinpointed us. It was wet, fully flared and quivering. Like the red eye of the Kanai elephant, the vision of the wet, quivering tip of the trunk has stayed with me ever since.

What happened next was beyond my wildest imagination. The animal got the fright of its life suddenly smelling a man less than three feet away. We at least had the advantage of expecting it; for the elephant it was a totally unexpected, shocking surprise. Without any vocalization, it wheeled around and tore back through the tall grass, smashing down everything in its way. We were able to follow its progress by its sound for quite some distance. It was so frightened that it had forgotten all about its friend. Almost reaching the next ridge it trumpeted a warning to its buddy. Then all was quiet. I turned round to Alo, and in a schoolmasterly way pronounced: 'Understand Alo, in this game, in the last analysis, it is nerves that matters.' Without a word, I shook Shingen by the elbow—a congratulatory handshake in the Garo style.

It was late afternoon and there was sunlight still. We decided that even if we had to spend another night out here in the adangs we must out-trick the makna. So far, if we spent a night to the right (east) of the road, it came out on the western side, reversing the order of its visitation matching our night watch. Readers might remember, Corbett also faced the same problem with the man-eating leopard of Rudrapayag. We did not have his option of closing the bridge across the river at night and confining the animal to one side of the river. We had only one option: to be on both sides of the road at night dividing the job between Alo on one side and myself on the other. Now that we were effectively two guns, we needed two assistants. A shikari needs someone to focus the torch correctly over the barrel of his rifle. For this we needed not only two powerful lights, but also two skilled men.

We decided therefore to send Shingen to Mahadeo about four miles from where we were and bring Ahmed, another powerful torch, and some food. Undoubtedly, thabalchu, boiled, roasted or fried and taken with salt and chilli, was replete with nutritional values, but after three days and two nights of it, our soul was crying out for a more familiar form of nutrition. We went down to the main road and Shingen left us near a culvert on the Rangara–Mahadeo road.

After Shingen left, Alo and I discussed our plans for the night meticulously. Alo with Shingen, the experienced local shikari, would go to Ramphachiring while Ahmed and I would spend the night in a crop-protection machan at Hatishia on the left of the road. The planning seemed to be logical. We sat waiting on the bench-like barriers of the culvert. Villagers returning home to Hatishia made polite enquiries of us. Minutes, then quarter hours, ticked by. The overripe day acquired a depressingly grey mould; yet, there was no sign of the jeep. Our impatience grew to anxiety. A few inquisitive stars were already twinkling mischievously in the sky. At last, when it was nearly dark, we heard the jeep coming up the slope from Mahadeo. There was time yet, then, to put our plan into action. But alas, Alo's wife Sumitra and my wife Sheila, their faces beaming, descended from the jeep along with Shingen and a tiffin carrier. It meant our well-laid plan had gone awry. I was too annoyed to feel pacified even by the sight of the tiffin-carrier.

We had been gone for two nights and three days without any news. Maheswari, an enterprising Marwari businessman occupying the room next to ours in the PWD shed in Mahadeo, had been most solicitous, a true friend of damsels in distress. As an old hand in the Garo Hills, a point he never hesitated to labour, he was very worried about our safety and had communicated his concern freely to the ladies. 'In this *bhari* forest, even the Garos get lost.' He repeated this message full of foreboding. Anxious to the point of distraction, my wife asked Ahmed not to sit idle in the bungalow but go and seek the sahibs out. Ahmed, gauging the mood of the ladies, did not dispute the point at that charged moment. He collected my shotgun, a couple of cartridges of birdshot, and stalked out of the bungalow. He was discovered a couple of hours later around a bend in the river fast asleep, his back resting against the dry river bank. When Shingen arrived with request for food and a torch, Sheila and Sumitra decided to prepare some luchis and aloo for their runaway husbands. That was the least, they felt, we would need after three days' privation. This naturally took some precious time; they, also naturally, did not realize how precious time was for us then. Heartlessly I had to make it clear that we had not needed luchis then, only Ahmed with a torch. But now instead of remaining with us, Ahmed would have to take them back to Mahadeo. That luchi and aloo nearly succeeded in making another notch in the long list of human victims of the makna.

After having the luchi and aloo and feeling, let me confess, all the better for it, we let the jeep go with our crestfallen wives and started trudging our weary way back. It was dark by now. One of the Hatishia villagers was returning home. I persuaded him to accompany us. He was quite enthusiastic when he learnt that one of us would stay at their machan that night. I climbed down a slope to an open-sided machan in the adang, which he claimed was his personal perch. Alo went forward to Ramphachiring with Shingen. We arranged to meet at the culvert in the morning.

Nothing happened on my side of the stage that night. A cold wind sweeping down the valley kept me awake, shivering uncomfortably through the night. April nights can be chilly in the Garo Hills. Early in the morning, I returned to the culvert. About an hour later, I saw Alo coming down the road in the company of a lot of jolly villagers, success written all over them like the banner headline of a tabloid. Yes, it was a success but nearly at a great cost.

Alo was walking to Ramphachiring with his torch switched on. Suddenly, in the beam of the light, he saw the huge back of an elephant just in front. The elephant had not sensed Alo's approach: so far it could only see its own shadow in front. Shingen and Alo stood a few minutes like that watching the tail of the animal swinging like a pendulum, waiting for it to move on. At that moment, they heard another elephant coming up from behind through the brush on the eastern side of the road. Obviously the maljuria pair was on the move, but there was no telling which was the tusker and which the makna. Shingen remarked, 'This is a bad spot.' 'Why?' asked Alo. 'Mosquitoes,' was the terse whispered reply. Even allowing for a Garo's reluctance to refer to the elephant by name, especially at night, this was certainly an understatement to beat all understatements. Shingen and Alo decided to yield the right of passage to the elephant closing up from behind, and scrambled up the eight-foot-high, perpendicular roadside cutting, holding on to the exposed roots of trees and plants, scraping for footholds on little jutting rocks. There was safety in the height.

In the dim starlight, they saw the elephant, the tusker, passing along the path to join its mate in front, its protruding spinal ridge floating past only feet away. They allowed the elephants half an hour and resumed the march to Ramphachiring, following the path we had taken earlier to come up to Hatishia in the morning. Climbing down the by-now-familiar bridle path to the adang, they sheltered in the old hut. At night, the makna came out into the crop fields. Alo, with Shingen holding the torch, approached it close and finally settled our account with the makna of Mahadeo.

Later in the morning when we took the forest staff from Phanda to inspect the carcass, we saw the footprints of the makna superimposed on the prints of Alo's hunting boots. Both had selected the same path to reach Ramphachiring—to their separate ends.

14

Identity Crisis
North Bengal, June, July 1975

The year 1975 was a turning point in my long association with elephants. I began to realize that the problem of man–elephant conflict was too complex for any short-term, single solution. The previous year had been a bad one for such conflict in North Bengal. The forest department hired a mela shikar team from Assam to capture young elephants, but it did not seem to be getting anywhere. The number of human deaths mounted to more than thirty that year. Then one year, after the figures went up to a staggering seventy, I embarked on a study. The result of my enquiry was a report for the forest department.*

The received opinion was that avalanches of elephants descended from the Bhutan hills in the crop season and caused the damage. My self-imposed task, the first ecological study of man–elephant conflict in India, was to map the paths of elephants and devise an advance-warning system about the course of their movement. I had no external financial support, no infrastructure but a hired jeep, a hired elephant and moral support from the West Bengal forest department. I stayed

---

*'Report to state government of West Bengal on the problem of elephant depredation in Jalpaiguri forest division and part of Madarihat range of Coochbehar forest division'. Calcutta, 1975.

in the old Sulkapara forest rest house for a whole month, monitoring the movement of elephants in one of the peak depredation seasons when maize was ripening in the fields. Tea gardens were the main locations of depredation. I realized soon enough that an average tea garden manager was more interested in keeping his files in order for claiming 'short-term loss' in his income tax returns, not verifiable on the ground after the lapse of months and seasons, than trying to contain conflict and prevent loss of human life. I recall a visit to the labour lines of the Kathalguri tea garden one morning to find five dead bodies lying on the ground in a row, all covered with white sheets. On another occasion while Dr R.K. Lahiri and Mr S. Palit, the Divisional Forest Officer of Jalpaiguri, were staying at Sulkapara, reports came in the morning of heavy depredation and human killing in Bandapani khas. These settlements consisted of widely separated clusters of family huts scattered over totally denuded forest land, most of which belonged to the Bandapani tea garden and the government. The hut-dwellers had encroached this land. When Dr Lahiri, Mr Palit and I reached the small family settlement in the wasteland, not a single hut was standing. Five persons, including one child, had been killed the previous night and one person had been seriously injured. There was a broken bird stand on the ground, a tattered volume of a book of alphabets, and footprints of elephants several feet deep ringing the cluster of huts. This seemed very odd. Why had the elephants attacked the huts? On being closely questioned, one of the survivors said that one or two elephants were trying to reach the harvested maize stuck in the rafters of the cottage, pushing their trunks through the windows. One of the inmates had taken a swipe at a trunk with a razor-sharp kookri. Very soon elephants encircled the cluster of huts. There was no escape for anyone. The herd systematically started demolishing the huts and getting at the dwellers within. It must have been an absolute nightmare in those desolate slopes below the Bhutan hills.

One of the main causes of the escalation of conflict was people encroaching the forests of the tea gardens, which was sometimes deliberately encouraged, it was said, by the tea garden management for a consideration. The government was considering taking over surplus tea garden land, and everyone was interested in paddling a finger in this no man's pie. Tea gardens themselves were trying to expand their activities to these 'waste' lands to prevent government from taking them over. The older topo-sheets of the 1920s show that these tracts used to be covered with dense mixed forest and actually served as a 'corridor' for elephants moving from one patch of government reserve forest to another.

An additional factor was that the southern part of the Rheti reserve forest had been acquired by the army to set up a corps headquarter at Binnaguri. This area in fact was a vital link in the corridor connecting the eastern part of the Central Dooars, from the river Torsa in the east to the river Jaldhaka on the west. It was also a key habitat pocket for elephants in the 'pinch' or dry season because of the availability of water in the area. This was a case of competition for a crucial natural resource, water, between man and elephant. Binnaguri was at the edge of the terai tract, whereas the other, northern, parts of Rheti were in the dry bhabar tract. With an India–China conflict threatening the region, the choice for the army was obvious. There was no difficulty in choosing priorities. Besides, the word 'corridor' had not yet become the buzzword among conservationists. The same thing happened to the Bengdubi forests of the Kurseong forest division, close to Bagdogra airport in North Bengal. After the army moved in, this soon became another hotspot of man–elephant conflict in North Bengal. An army internal report on elephant damage in the Bengdubi Army Supply Depot acknowledged that elephant depredation in the area started with the establishment of the army depot which was encroaching on the elephants' territory. Again, the availability of water had dictated the choice of site.

My month-long stay at Sulkapara taught me to live with and share the concerns of the community. Every evening a procession of people moved across the paddy fields from Grassmore village in the Central Diana reserve forest to the Sulkapara marketplace for security from elephants. A patch of riverine forest, the Central Diana was where a herd of more than fifty animals chased from the east by mela shikar parties had taken refuge. The brunt was being borne by the villages around Diana.

In the Central or Western Dooars, between the rivers Torsa and Jaldhaka, I could not locate any point where elephants came down from the Bhutan hills. The forest department had prepared a flow chart of the movement of elephants in the Western Dooars, i.e., the Jalpaiguri Dooars, excluding Buxa in the east. Their hypothesis was that the elephants poured down from Bhutan through the Titi reserve forest bordering Bhutan, where Titi and the Western Dooars were contiguous. I had worked systematically from Jaldhaka eastward, but had failed to detect any point of ingress from Bhutan. In fact at the Jitti-Hope tea garden right on the border of Bhutan, near the gorge of the river, they had never seen wild elephants. The northern boundary of the Titi reserve forest, west of the river Torsa, stretching down to Jaldapara wildlife sanctuary remained the last point of mystery; it was already recognized as the main area of man–elephant conflict in the Jalpaiguri Dooars.

Mr Subimal Roy dropped in one day at Sulkapara. Earlier Divisional Forest Officer of Jalpaiguri, he was now Divisional Forest Officer of the adjacent Coochbehar forest division which managed Titi as well as Jaldapara; he was naturally very keen to probe the mystery of Titi. We agreed that the only way to make sure that Titi was the entry point for the Bhutan elephants into Jalpaiguri was to traverse on foot the entire northern boundary of Titi, where Titi abuts on Bhutan, and look for well-used tracks. Accordingly, we set out for the western edge of Titi bordering the Lankapara tea garden. It was quite a walk from the western to the eastern end. Mr Roy had arranged for a young departmental tusker, Lal Bahadur, to carry us across. Lal Bahadur, who grew up to be a grand tusker, was always a bit nervy; however, at that time he was not yet in musth, that is, not mature yet. Later, in 1976, I was on Lal Bahadur when taking a census of elephants in Buxa. He bolted through thick forest when a herd with a large female elephant approached us to satisfy its curiosity. I managed to save my head by flinging myself down flat on my back on the gaddy while stout branches swept past overhead. My hat was knocked away, but fortunately, my head was spared. Lal Bahadur ran through dense thickets of cane, reducing my clothes to tatters and deeply scarring my body and face—one ear sustaining a particularly nasty gash. When the wild stampede ended and the animal stopped trembling, I returned and retrieved my hat. Another bolting followed a quarter of an hour later, with more damage to my skin and clothes.

But at that time I knew nothing of Lal Bahadur's foibles. We had to climb about a hundred feet or so to reach the boundary track between Titi and Bhutan, which had been knocked through the jungle by the forest department. Young Lal Bahadur made so much fuss going up the hill, breaking wind resoundingly and squealing in protest, that we decided to abandon ship, i.e., Lal Bahadur, at that point and footslog the rest of the way to Ballarguri, a village on the eastern limit of the forest. It was a charming path, densely wooded and lonely, with a close canopy of trees. The ground was fairly level. There was no sign of any elephant track crossing from north to south across the path. There were no human tracks either.

At one point, we met a loquacious villager coming from Ballarguri and Totapara on his way to meet friends and relatives in Lankapara. As politeness demanded, he enquired what we were doing there. In five minutes or less he then outlined the movement path of elephants in that stretch. They came from farther east, he said, went along the foothills to the west, and after a brief visit to a saltlick in a place called Sinkawle Pakha on the Indo-Bhutan border, just north of Rheti, moved on to the Rheti forests; he neatly summarized the information we had been gathering over the weeks. Mr Roy and I burst out laughing, admiring the simplicity of his explanation. Didn't the elephants come to Titi from Bhutan? we enquired. No. There was a table-like land formation in Bhutan just north of Titi. The forests in that part of Bhutan had all been converted to orange orchards. There was no elephant north of Titi. Our pet theory of elephants debouching from Bhutan went bust at once.

In a continuation of the investigation the next autumn, I realized that east-of-Torsa and west-of-Torsa were two separate populations, their ranges periodically extending and overlapping around the Torsa. The customary practice of the forest department till then had been to allow capture of elephants in a kheda-cum-mela method around

Kurul in the north- eastern corner of Buxa on the Bhutan border to contain man–elephant conflict in the Western Dooars, west of the Torsa, without realizing that these were two different sets of animals. The Jalpaiguri elephant population was made up of two practically separate populations. One blanket prescription could not cover both. If you had a sore head, you did not seek relief by applying balms to the small toe in the foot.

This caused some embarrassment to the forest department. Mr B.C. Ghosh, a tea magnate in the Western Dooars, was prone to go to court at the drop of a hat. He had sued the forest department for damage caused by elephants to his Tondu-Bamundanga tea garden. The government pleader had taken the line that as the animals concerned were Bhutan elephants, the West Bengal forest department was not responsible for them. There was an ancient agreement between India and Bhutan similar to one between Assam and Arunachal Pradesh, that a part of the royalty received from the capture of elephants in India would go to Bhutan, as they were the same elephants crossing the international border at will. Of course, things had changed since then but this did not make the legal stand of the government pleader any the less embarrassing. And now, to top it all, it appeared that the elephants had no connection to Bhutan at all.

II

Apart from a study of elephant routes and corridors, my other task was tackling two declared rogue elephants, both maknas. One incident is still etched in my memory.

The Tondu-Bamundanga tea garden, a few miles south of Sulkapara and just east of the Garumara wildlife sanctuary across the river Jaldhaka, was receiving a lot of unwelcome attention from wild elephants. This being one of Mr B.C. Ghosh's gardens, with claims for damage to garden property by wild elephants already before the court, I was informally asked to look into the matter. One of my local friends, Goresh Ali, and I, went over in my jeep and checked the reports. Next day late in the afternoon, we returned to the tea garden. Receiving notice in advance of our programme, the manager had gathered some garden workers to act as our local guides. They said that the elephant came out of the South Diana forest every evening and raided the labour quarters, their crop, and garden property. The staff showed me the broken-down walls of the brand-new hospital. I asked them, to make sure, if it was the makna. They said yes, unhesitatingly. Admittedly, there was some confusion when someone from the back row, to emphasize the size of the animal, mentioned its big tusks. Soon such irrelevancies were suppressed. We were looking for a makna of the size mentioned in the official order. Of course the mere size of the footprint is not enough; there must be, at any given time, dozens of elephants with footprints of approximately the same size. The system has been fine-tuned since then and some other identifying marks, such as the shape of the back or a description of the tail, are also mentioned now in notifications to avoid wrong identification.

They told us that every evening the animal walked across the garden's football field along its southern border. Accordingly, with Goresh and the local guides I took position on the border of the football field. It was the last week of June. There were spurts of not-so-gentle drizzles and a watery moon was drifting through strips of monsoon cloud. Soon there were furious barks from dogs in the labour lines on the other side of the ground. The moving wave of the din clearly indicated the line of the elephant's progress. Our local guides also noted this. Knowing the area well, they said the animal was moving towards a low grassy area beyond the football field—apparently one of its favourite grazing grounds in the garden. They said the area could be approached by jeep along one of the tea garden roads. Heeding their advice, I came back to the jeep and reached the motorable path on the other side of the field. The drizzle had stopped and the moonlight was brighter, but fleeting clouds were chasing the moon, plunging us into darkness now and then. As we crept forward, a hundred yards or so down the path I could hear the elephant feeding in the grass. We stopped the jeep and Goresh and I got off. We moved forward carefully. The sound of the feeding on grass was very clear now. The path ended abruptly at a depression and the elephant was in it, feeding on grass and *Alpinia*. I approached close to the rim of the depression. The elephant was just in front of us. I could see its grey form in the dim moonlight, but as it was standing in a slightly quartering-away position, I could not see its head and therefore could not be sure if it was a makna or a tusker. I had learnt by then that one could not trust hard-pressed villagers' description in such matters. For them it was a struggle for survival. They would happily affirm whatever description the sahib suggested and would agree to it being a four-tusked elephant, two growing in front and two at the back, if that satisfied the sahib. Keeping the rifle in the ready, I asked Goresh to switch on his light. The elephant turned round slowly and, far from being tusk-less, it revealed a big pair of tusks. It took a few enquiring steps forward. We ran for all we were worth to the jeep, the tusker coming up from the depression in the ground to the garden road to see us off. The driver had had the instinctive good sense to turn the vehicle around. Perhaps he had anticipated the sudden need for instant flight. When piling into the jeep, I had a last look back. The grey form was silently walking up the path.

We reported the evening's proceeding to the garden's managerial staff and explained to them that it was not a makna but a big tusker, and out of bounds for us. They understood my point but looked extremely unhappy.

Sulkapara was a mere 45-minute drive away, but that would involve crossing the Tondu river, a tributary to the Jaldhaka, on the way. The ferryman would hardly be available at that hour of the night. They offered to put us up for the night in the garden's guesthouse. At about one o'clock in the morning, there were loud thumps on the door and shouts. The elephant was back in one of the labour lines causing havoc. Would I come to their help? I knew I could do nothing. The terms of the order clearly specified the offending animal to be a makna. I had seen with my own eyes that it was a tusker. But maintaining good public relations is important in such a situation. I said, of course I'd come and see what I could do, knowing very well that there was precious little I could actually do. But at least they would not feel abandoned and let down at a time of extreme stress.

We reached the No. 12 labour line. The drizzle fortunately had ceased. We stopped the jeep a few hundred yards short of the labour line, as instructed by our guides, and followed them on foot. There were two rows of huts on either side of a narrow lane, and huts on the far side of the path lining the South Diana reserve forest. I could hear the elephant. We stopped behind a hut on the near side of the line and peeped through the narrow passage between two huts. Across the lane I could see the tusker raiding the crop grown by the labourers in small plots in front of their huts. I watched the animal from a safe distance of about thirty feet. There was no sign of life in any of the huts—they had all been abandoned, apparently, for the time being. The tusker demolished the crop in a small plot, and then stepped forward and put its tusks under the eaves of the hut and jerked it upward. It flew open like the lid of a box. The animal lingered around for a few minutes, presumably exploring its inside for more edibles. I could not see from the rear exactly what it was doing, but this seemed the likeliest scenario.

For two hours I followed the stately progress of the animal down that narrow lane as it systematically repeated the process in every hut. For some reason it concentrated its efforts on the line of huts bordering the forest, not turning on the huts behind in which we were sheltering. All the time I was carrying my loaded rifle more as a gesture of goodwill towards the workers than as an effective deterrent. At the first hint of dawn, the elephant melted away in the Diana forests.

Returning to Sulkapara in the morning, I sent a written message to the Divisional Forest Officer that this tusker's propensities certainly made it a threat to human life and a potential candidate for declaration as a rogue.

III

A common night-time experience of visitors to Garumara on the border of Diana was to find a huge, old makna blocking the way. One either had to turn back or wait for it to move off the road when it finally chose to. It was also a frequent visitor to the saltlick that the forest rest house overlooks. Any time after two in the afternoon the makna began his wandering. When I went through the order declaring it a rogue, I saw that going by the measurement of the footprint of the forefeet its height was around 10'2". Though the North Bengal forests have large numbers of massive male elephants, this size was

unusual. The description 'makna' narrowed down the search further.
I was sure the order referred to the old Garumara makna, which
I too had encountered in the past, a gigantic barrier across the
Garumara road.

From the list of its misdeeds attached to the declaration order, I
found it had killed a whole family a few days before at a village called
Dhupjhora on the northern fringe of the old sanctuary boundary.
Only one child and its mother had survived.

It then crossed the Jaldhaka river and was currently hanging
around Khairbari, about two miles from Sulkapara. Just before my
arrival, it had chased and seriously injured one person at Khairbari.
Mr Palit, the Divisional Forest Officer, had gone there atop a
departmental elephant for a spot-inspection. His mount displayed
distinct signs of agitation on reaching the spot. Both Mr Palit and
the mahout agreed that the killer makna was probably close enough
for the departmental animal to smell its presence. Mr Palit, being
unarmed, the party wisely retreated from the area.

A few days after I had settled in at Sulkapara, one morning, a
group of grim-faced villagers appeared, accompanied by the local
forester. They had bad news. Early in the morning, a villager at
Khairbari had heard some slight noise from his field of maize.
Suspecting thieves at work, he climbed down from his cottage built
on stilts and went silently into the field of maize to catch the miscreant
red-handed. Instead, he found the makna, breaking cobs of ripe corn,
as stealthily as a thief. Our villager tried to flee back to his cottage,
but the elephant noticed and padded steadily behind him. He was
able to reach his cottage, but was pulled down when he was already
halfway up its wooden ladder.

We clambered up on to the elephant Malati's generous back, a
party of three, besides the mahout. The forests of Diana at the end of
June were a network of swollen streams. An elephant was the only
possible mode of locomotion in that riverine tract of khair and sissoo.

Leaving the dead body on the ground of the hut surrounded by
wailing family and friends, we started spooring at once. This was the
only time I stalked a rogue on elephant back. I made a routine check
of the diameter of the footprint, very prominent on the wet ground

and in the crushed undergrowth. Malati was quite unperturbed by the proximity of the rogue: she only lifted her trunk now and then and pointed in the direction she could smell the rogue.

We had not proceeded even half a mile from the hut into the forest when I spotted from the elevated position of the elephant's back the supine forms of *two* elephants. Had we been on foot, the undergrowth would have prevented me from seeing them as they were lying down. They were only about fifteen feet away. I was not sure of the correct angle of a fatal shot from a high position. I was getting ready to slip off Malati's back. Just then Malati decided to flap her ears loudly. Both the elephants were up like a shot—they were both maknas. Both were large animals, but one was so much larger than the other that I knew at once which was my quarry. The larger animal was facing us straight with its ears flared, one front leg swinging back and forth; the other elephant, a relatively smaller animal, nearer us, was walking across our line of sight, about to obscure the view of the larger animal readying itself to launch on a charge. That animal obviously was the immediate threat. I fired at the head of the larger animal straight ahead. It turned to its left and ran. Indifferent to the sound of rifle fire, the smaller one did not alter its course—it went on closing in on us. Even a push from the animal, which was just a few feet from our mount by then, would have flung some of us off Malati's back. It was a critical moment. I could not allow it to drift any closer. I could see its temporal gland in full flow on the near side, wetting the cheek down to its jaw. The strong odour of the musth fluid was overpowering. I dropped it in its tracks with a single temple shot. Meanwhile, the larger animal was making squeaking calls from a point some fifty yards away. I clutched the shoulder of the mahout to point out which way he should proceed. His body was in an uncontrollable fit of shakes— *hathi jokar* as it is called—and he sat frozen, unable to respond. I had to neutralize the effect of his fear-induced shock with a serious counter-threat. I jabbed him from behind with my rifle muzzle and threatened to shoot him on the spot if he did not move forward. We proceeded following the track of the larger makna which had just fled. About fifty yards away, it was standing sideways behind some khair bushes. It started turning towards us hearing our approach. A couple of more shots, somewhat better placed now, saw its end.

Some elephant ecologists have claimed that single males forage together at night and rest separately during the day. This was also my experience at Rosekandi and Bengdubi. Here was an ocular demonstration that the reverse could also be the case. There was an embarrassing sequel to this. The declaration order specified 5′1″ as the circumference of the front foot; an additional mark of identification was a large burn-mark on the makna's back. I found only the smaller animal bore the burn mark, not the larger one. Both being maknas, the description of the two animals, observed at night in torchlight, had got mixed up: the height belonged to one animal and the prominent burn mark to the other. Such burn marks are common among marauding elephants in North Bengal. Tea garden managers supply burnt engine oil to their labourers to prepare flaming torches to keep away marauding animals. Some of the labourers pushed to the extreme are tempted to throw large oil-soaked burning rags on elephants. The rag soaked in oil sticks to the back of the animal. The result is a spectrally illuminated elephant running through tea bushes and the labour lines, much to the general amusement and satisfaction of all the aggrieved. This was yet another way to create a rogue elephant. The responsibility of the tea garden management ended with the supply of oil to the labourers. Supervising its proper use was not their concern. Diesel purchased in the open market is regularly supplied to villagers in southern West Bengal in the depredation season, for the same purpose, not always under the supervision and control of the forest staff. Neither the labour unions in the North Bengal tea gardens, nor the village panchayats or forest protection committees in southern West Bengal insist on a proper use of the oil supplied. Nobody is interested in the problem once the animal has been chased off his patch of land.

15

Teenage Aggression
North Bengal, July 1988

I t is a common misconception that only mature, large male elephants, especially the solitary ones, are aggressive and are likely to threaten human life. Records and my own experience speak differently.*

In October 1968, a young sub-adult tusker was wreaking havoc in the Garo Hills in the forests around Rongchugre on the Phulbari–Tura hill road. It was hanging around on the periphery of a herd of about twenty elephants permanently resident in the area, and had killed five persons before it was eliminated as a rogue. The height of the animal, calculated by the twice-the-circumference-of-the-front-food method, was only 7'2". (The height of an elephant at the withers, barring that of calves and very old animals with splayed toes, is calculated at twice of the circumference of the forefoot with a margin of error of c. 2 per cent. The average minimum height of an adult male elephant is 8 feet or 245 cm, and that of a female elephant 7 feet or 215 cm.)

*The method of capture and immobilization described in this chapter closely follows the method I witnessed in Malaysia in 1988. See 'Translocating Elephants,' *India Magazine*, January 1990. Plates 15 and 17, and the cover photograph capture my experience of this method in Malaysia.

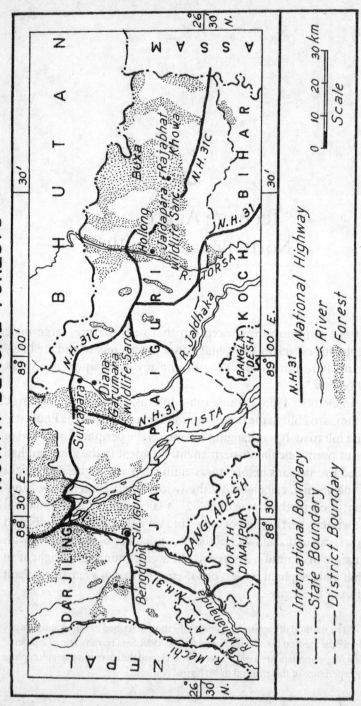

North Bengal forests

In October 1969, a herd of elephants was rampaging around the village Chigichagri in the Garo Hills, east of the Fulbari–Tura hill road, a few miles southeast of the Rongmachak PWD inspection bungalow. As requested by the top district administration, two elephant control licensees destroyed the 'leader' of the herd in an attempt to scare away the animals. While stalking the herd, the hunters received a determined charge from a makna barely seven feet tall. The recalcitrant animal pressed on with the charge even after being hit on the head frontally, and eventually fell to two solid bullets in the head, only a dozen paces from the hunters.

In November 1985, I was attending a meeting of the Asian Elephant Specialist Group of the International Union for Conservation of Nature, Survival Service Commission, at Bandipur in Karnataka. Late one afternoon I slipped out of the Bandipur tourist and administrative complex with J.C. Daniel of the Bombay Natural History Society, in the Society's jeep, looking for a lame elephant which had been reported to be lurking nearby. Close to the gate, on the national highway, we encountered a young tusker which insisted on disputing our right of way and demonstrated at the jeep with 'mock' charges. The rush began with the head in a low position, and ended in an upward thrust of the head. We sat out the rush in the jeep, and the charge was not pressed home. The dividing line, be it known, between a 'mock' charge and a serious one is very thin; the proof of the pudding, if one may mix metaphors, being often in the killing by the animal.

Next year, again, in nearby Mudumalai, a small tusker barely 7 feet or so at the shoulder, made determined charges at our jeep. Dr V. Krishnamurthy, who was accompanying us, said that there were two such animals of the same height lurking in the area, one a 'left' tusker and the other a 'right' tusker, and both of them were notorious for charging at vehicles. He was worried that one day the animals would damage a vehicle and injure its occupants, who were usually tourists.

In April 1986, census parties of the West Bengal Forest Directorate moved on elephant back through the forests of the Jaldapara wildlife sanctuary in North Bengal counting wild elephants. As they searched the forests, the departmental tusker Lal Bahadur (height c. 9 feet; since deceased) carrying a census party was determinedly charged by a young chakna tusker well below 8 feet in height, whereupon, true to form, Lal Bahadur fled. This wild young bull was notorious in the area for its aggressive behaviour, though it had not killed a man yet.

On 26 November 1986, a small tusker was declared a rogue for killing a person in the Lohar Singh division of the Mary View tea garden in the Kurseong forest division in North Bengal, and was shot dead on 7 January 1987. It had killed in all five persons and injured one. The circumference of the front foot of the animal was 3'10$^{1/4}$" (1.19 metres), and therefore, the height of the animal at the withers would be around 7'9$^{1/2}$" (c. 2.38 metres).

On 7 January 1987, a young makna (circumference of forefoot 99 cm; estimated height 1.98 metres or 6'6") killed one Ukil Roy at Tista Bhanga Ghat near Nodhabari village in North Bengal at about 11 p.m., chasing the man over a distance of 200 metres.

One evening in 1987 in the proposed Royal Manas National Park in Bhutan the Chief Warden P.B. Subba was out patrolling when his jeep was attacked by a sub-adult tusker and smashed beyond retrieval. By some miracle P.B. Subba escaped alive.

On 30 May 1988, at about 3-30 a.m., an army jawan, one Dayaram Naik, was killed by a tusker near a gurudwara within the army campus at Bengdubi in North Bengal. He was sleeping on the veranda of the gurudwara along with two other persons when a small tusker in the company of a larger makna came close to where they were. Dayaram tried to run away. The tusker chased him over a distance of 50 metres, held him by the neck with its trunk and killed him— an unusual way of killing by an elephant. The measurement of the front footprint was 122.5 cm, which means the approximate shoulder height was 8 feet.

In May, June and July, this animal seriously injured three other persons; one of the injured, an army jawan, reportedly later suc- cumbed to his injuries. I was told that a lady, pillion-riding behind her NCO husband on a scooter, was picked up by this animal, which ambushed them from behind. She got away with only minor injuries

as the animal dropped her and beat a retreat, frightened by the ensu-
ing fracas to which, no doubt, the *phutt-phutt* of the scooter engine
had contributed generously. The commonly held view around the
army camp, however, was that she had been spared in response to
the husband's earnest prayer to Ganesh Baba, offered with folded
hands and due humility.

Be that as it may, on 7 June the tusker struck again. At about
9.30 in the evening, it approached one of the labour huts of the
Belgachi tea garden in the Kurseong subdivision of Darjeeling civil
district, only a few miles from where Dayaram Naik had been killed.
The hut was occupied by one Saligram Karua, his wife Budhani,
and their two children, who tried to run away. The unfortunate
who did not make it was Budhani. She was chased over a distance
of about 30 metres and trampled as well as gored to death. The
measurement of the circumference of the front footprint of the
animal was 122 cm, close enough to the footprint measurement of
the killer of Dayaram. Besides, the description of the animal on the
two occasions, gathered locally, also tallied.

Aggression in adolescent/immature human males has been
widely debated and seriously studied; but such behaviour in elephants
does not seem to have received adequate attention yet. These animals,
as in the case of men, often are a nuisance, and seem to delight in
proving their might to themselves and to the wide world by chasing
people, not infrequently to the bitter end.

It is tempting to hypothesize on the cause of such behaviour. Is
it due to difficulties in social adjustment in a normally gregarious
animal, soon after expulsion from the family group when nearing
the age of sexual maturity? One notes how some of these expelled
sub-adult animals continue to hang around the herd/family group,
moving along with it at the time of migration; and how some of
them, probably at a later stage, graduate from this hanger-on status
to the role of junior partner in maljuria groups. This way the process
of learning by observation and emulation continues even after
expulsion from the family group.

Such aggression may also have implications for the reproductive
behaviour of elephants. Take the case of the Jaldapara chakna, referred

to earlier. It became a nightly visitor to the Hollong pilkhana of the sanctuary, where it had managed, brooking no competition, to mate with almost all the available departmental cow-elephants, going to the extent of causing an abortion in a gravid cow by forcing its unwanted attentions on the expecting mother. Another time it unhesitatingly took on and gored Balaji, a young but very tall (for its age) departmental makna (slightly over 9 feet or 275.8 cm, since deceased—a foot taller than the chakna). It seems possible, therefore, that body size alone is not the decisive factor in competitions for mates: the meek, even if large and hunky, not necessarily inheriting the cow.

The management aspect of the problem poses some teasers. The present practice is to tolerate the aggressive behaviour up to a point; and when it hardens into man killing, to destroy the offender. With such animals, three other options are theoretically open to managers: (i) chemical immobilization and translocation to a safer area (practised successfully with a rogue tusker—8'9"—by the Karnataka forest department in March 1987); (ii) capture by chemical immobilization and subsequent training; and (iii) immobilization/ sedation, followed by a sound thrashing by larger koonkis, imparting to the youngster a never-to-be-forgotten lesson in good manners, and eventual release.

With the first option, it is desirable, experimentally, that a nylon collar, preferably with luminous paint for easy detection at night, be put on the animal before release, so that the behaviour and movement of the animal can be monitored later. A radio-collar, of course, would be the ideal thing, but this would require proper equipment and trained staff, not always readily available to field managers. Capture by chemical immobilization followed by training (option ii) has been attempted successfully with young bulls. It was tried, for example, with a young tusker, later named Agasthi, by the late S.R. Choudhury of Orissa, and with a sizeable number of elephants at Kattepura by the Karnataka forest department. Most of these animals, however, were not known to be especially aggressive. It has been seen in Karnataka and West Bengal that even aggressive animals respond well to the training process. Incidentally, the time may not be far off when mild chemical sedation will become standard prescription in the initial stages of the training programme for recently captured wild elephants, when the stress and strain of getting used to constant human proximity is at its greatest. As for an aggressive animal's response to the training process, Janardan, the calf of a forest department cow elephant, Jaymala, in North Bengal, was extremely aggressive at the age of three. The animal, nearing 9 feet now, though still spirited, is after training disciplined and tractable. Till the first quarter of 2002, Karnataka's forest department had captured and trained fifty-five adult male elephants with few casualties. A vicious killer tusker, captured, trained, and named Rupnarayan, became perfectly tractable, but unfortunately died two years later from other causes.

The experiment with thrashing by larger koonkis (option iii) seems worth trying out, as the dominant position of larger animals is usually accepted by the smaller and younger ones. This is the role the *sardari koonkis* ('leader koonkis', a term used in north-eastern India: usually tuskers of outstanding size and bulk) play when bringing freshly captured wild elephants out from a stockade.

In the last week of July 1988, the West Bengal Forest Directorate had to choose between the available management options for dealing with the man-killing young bull of Bengdubi. The option of liquidation was relegated to the background, but not closed. It was decided that the other available options would be tried out first.

Initially, since the animal was not unmanageably large for the available departmental koonkis, both the first option (capture, translocation and release) and the second (capture and training) were kept in view. The argument in favour of Option 2 was that in the conditions prevailing in North Bengal one could not rule out the possibility of further conflict with man subsequent to the release of the elephant. The argument in favour of the first option was that the animal, if released in a reasonably large area of natural forest with a low density of human population, would be free from the stress of living in small patches of degraded forest hemmed in by human settlements and in daily confrontation with man. It might eventually rehabilitate itself in its new surroundings without continuing to pose a serious threat to human life. A nylon collar with luminous paint would be put on the elephant, if possible, the first thing after capture. It would therefore be possible to identify and recapture or eliminate the animal without much fuss if it continued to be difficult even after its release. However, this business of the nylon collar was forgotten at the time of the actual Bengdubi operation.

Experienced trainers from Assam cautioned that the chance of mortality in the course of training this particular elephant (height 8 feet) in the hot, humid weather of July would be unacceptably high. Besides, it would be better to try and rehabilitate the animal in the wild than trying to domesticate it. On these considerations the forest department finally decided to exercise the first option (capture, translocation and release). What followed was a chain of accidents, disasters and near-disasters which no one could have foreseen.

The forest department meticulously planned and made all preparations. A team of koonkis was especially brought from Jaldapara, all of them except Urbashi, alas, now dead. The team included Jatra Prasad, Chandrachud, Balaji, Rajkumari and Urbashi. A departmental lorry with a suitably modified body arrived at the same time. A temporary forest camp was set up and a ramp following the Malaysian design was built inside the Bengdubi army supply depot. A similar ramp would have to be built when unloading the captured animal from the lorry at the spot selected for release deep inside the forest: in this case the heart of the core area of the Buxa tiger reserve near the eastern part of the North Bengal forests, not too far from the border of Assam. Training sessions in loading and unloading the captured animal from the lorry were arranged for the koonkis and mahouts. It was an exercise new to all: the forest staff, the mahouts, and the koonkis. It was as much for training as for confidence-building to face a potentially dangerous situation. A big bandobust was slowly building up. The services of P. Guha, a renowned local hunter and tea garden manager, were requisitioned to provide heavy-rifle cover during the operation. I was invited to participate in planning and executing the operation as I had actually seen such an operation being carried out in Peninsular Malaysia only a few months before.

Hiccups started even before the word 'go'. I was down with a prolonged bout of debilitating slow fever, but, obviously, the call of the elephant could not be ignored. Such high-profile bandobusts attract a lot of media attention however and it just would not do to get myself reported as chasing killer elephants while I was actually on long medical leave from my place of work. My personal doctors were most accommodating, but my wife insisted on accompanying me. Beware the press, not to speak of the wife, became my motto throughout the trip.

The working team spent the first day on elephant back doing a dekko around the army camp, which had been set up after the brush with the Chinese in the early 1960s by clearing large patches of fine evergreen reserve forest. The party returned late in the morning and reported that they had seen the chakna, but it was in a maljuria group of four, one of them a huge, almost rotund, makna. The same evening my wife and I had an occasion to meet this worthy, but more of that later. As soon as the darting team on elephant back approached the group, they reported, it moved off, denying the party the chance of a tranquillizing shot. This happened several times. Eventually they decided to give up the chase for the day as the koonkis had to be rested and fed.

Late in the afternoon as we were sitting on the first-floor veranda of the Bengdubi forest rest house, we saw groups of people hurrying eastward towards Bagdogra, some even in cycle-rickshaws. Curious, I asked the chowkidar the reason for this sudden one-way movement of people. His casual reply was that they were going to see *the* elephant. Apparently, this elephant appeared on the army golf course every day late in the afternoon and then, close to sundown, moved to the army hospital compound to feed on the succulent grass there. The curious flocked there in time to see the daily show, but not our blasé chowkidar who considered himself above such trivialities. We shouted for our jeep and proceeded towards the golf course, less than a mile from the bungalow.

The golf course extended south from the road to the forest, and sure enough we could see a very large makna standing on the edge of the forest beyond the fairway. A small crowd had already gathered on a side road on the eastern boundary of the golf link. Breaching all club etiquette we drove across the fairway to get close to the elephant for a good view. Only a narrow stream separated the forest from the southern edge of the golf course. The makna was a really well-rounded specimen. Casually picking up titbits *en passant*, it was slowly ambling towards the stream, that is, towards our jeep, stopping frequently to cast an inquisitive look at us. Then a cyclist, obviously an old-timer, came up from behind, and asked us to move on as our vehicle was standing squarely on the elephant's regular track. His main concern appeared to be the elephant's convenience, not our safety. The animal itself was barely fifty metres from us now. We hastily drove forward to the narrow side road at the eastern end of the golf links to get out of the elephant's way. The sun's rim dipped further, the elephant ambled out of the forest edge, and in easy casual strides—no undignified rushing out for him—crossed the stream and got on to the side road where we were, less than fifteen metres from our jeep. Its head was turned towards the main road to Bagdogra and the hospital compound beyond. The assembled crowd provided vociferous opinions on the situation. Some knowledgeably claimed that it was a domesticated animal from Nepal, now turned feral. So indifferent to human presence was the animal that the hypothesis seemed perfectly plausible at that moment. It was proceeding calmly with its regular evening routine to the hospital compound across the Bagdogra–Bengdubi road, or so the people said. It did look as if it had a pre-selected destination in mind. The pace was now brisker and the animal had the purposeful air of one slightly late for an appointment.

What followed was as bizarre a spectacle as one could wish for. The side road was a narrow, metalled strip. The fat makna was apparently making straight for its ultimate destination: the hospital compound. We were following its stately progress in our jeep, accompanied by an animated crowd of about two hundred, a quarter of it around the age group of ten; the phalanx was headed by the makna. The traffic was by no means one-way. People were coming from the other direction as well. They pulled to the narrow, muddy shoulder of the road allowing the makna the right of way. It passed within feet of the excited crowd. The cycle-rickshaws coming towards us followed the same drill, the passengers not even bothering to get off the rickshaws. The makna continued to push along, almost brushing past the rickshaws. At this stage some urchins started throwing stones at the stately animal, temporarily disturbing the smooth rhythm of its progress. It stopped and turned halfway round as a threat gesture to the unruly and disrespectful young. It was really more a gesture of indignation than of threat, nothing serious: at least nobody took it as anything else. At last the procession reached the Bagdogra–Bengdubi road. Beyond lay the medical centre and the surrounding fields, unkempt and very sylvan.

There was a lesson in this experience. Contrary to what some naturalists have claimed to have observed, I have repeatedly seen male groups gathering together during the day, with their members going their several ways in the evening on their foraging or marauding expeditions. That very morning our darting party had come across the same makna unmistakably identifiable by its huge bulk in the male group of four which included the rogue chakna.

The next day's experience was grim. The darting party had been after the male group from the early hours of the morning till well past midday when the hunt was called off to feed and rest the koonkis and men. Unfortunately, the Chief Conservator of Forests had decided to visit the camp the same day to see how things were going. A forest range officer had been specifically detailed to arrange for fodder for the koonkis on duty as they would not have time to collect their own; but in the excitement of the chief's visit the range officer had neglected his primary assignment. So, after a hard morning's labour, the koonkis found themselves tethered down in their stalls, without any fodder. Rajkumari, a sedate and affectionate elderly female koonki, was standing in her stall next to Balaji's. Kartik, Rajkumari's patawallah, was squatting down in front of her, making bundles of grass for his charge. Balaji was still waiting to be served. Tethered, with his mahout still astride his neck, he had no way of helping himself to the surrounding greenery. We had a hungry and angry Balaji on our hands. Suddenly Balaji lunged sideways and grabbed Kartik by his trunk and started striking him against his legs.

Kartik screamed; cries of horror rose from all the elephant men around. Grabbing their elephant spears, they rushed to the scene. Balaji's mahout applied his *ankush* (elephant goad) on the elephant's head with all his might, screaming all the while at his mount to let go of the unfortunate man. Rajkumari was trying her best to break her chain to come to Kartik's rescue, but, unfortunately, the chain held. All the mahouts and patawallahs assured me later that if only Rajkumari had been free and not chained down, she would never have allowed Balaji to kill. When Balaji eventually released Kartik, he was still alive. We rushed him to the large army hospital where his last words were: 'I could never make Balaji accept me.' Balaji himself was gored to death by a wild tusker in Hollong camp a few years later.

We had a hasty meeting with all the elephant handlers and asked them if they were still willing to continue with the exercise. As one man they replied that death was an occupational hazard which they had come to accept. I suddenly realized that though I had known these men for so long, I was just an outsider when it came to the crunch: their professional ethics and value system. I could only admire it from a distance but could not intrude upon their inner world.

D-day minus 1: Next morning the darting party started out again after the male group which included the killer chakna. They located the group early enough in the army supply depot campus. As usual, when the koonkis approached, the group started drifting away, never in a hurry but always exasperatingly beyond the range of the dart guns. I accompanied senior officers of the forest department in a jeep to follow the movement of the koonkis. There was no point in non-technical persons without specific job assignments using up the limited seating space on the koonkis.

Some time past midday we met the darting party on a road within the army campus, and asked them to call it a day. The experience of the temper of hungry male elephants the previous day was still haunting us. Kartik was no longer with us, but Balaji was. The operation needed the help of large koonkis; besides, a well-fed Balaji was perfectly tractable. Mr Guha soon left with his heavy rifle for his tea garden, not too far away. All of us, not only the koonkis, were thinking of food. The koonkis were dismissed and they immediately left for camp on the double.

Just then a passing NCO on a bicycle told us that the chakna was close by. A hasty roadside conference followed. There were two schools of thought. Some favoured immediate following up of the hot trail; some did not. No koonkis, no Mr Guha, no heavy rifle, why not wait for the next day? The elephants were not going anywhere: they had obviously selected the sprawling army campus as their favourite daytime resting place. A few army men were around to be sure, but no disturbance from the noisy, teeming *hoi polloi*. The Ayes argued that they had tracked wild elephants on foot safely in the past without any heavy-rifle cover. The party, they argued, had been moving for some days now on koonkis without success. Why not try on foot now? The Nays countered that while there may not have been any accidents in the past, there was always a first time. The Ayes ultimately carried the day and the party went forward on foot led by the local field officer S. Dhandiyal, and accompanied by another officer, Pradeep Vyas. We were to await them in our jeep at a pre-arranged spot.

Frankly, we, the seniors, were treating it as an ineffectual exercise which, nevertheless, had to be gone through for the sake of the morale of the troops. We waited on the road to the forest bungalow, just as a matter of form, looking impatiently at our watches every few minutes for the party of adventurers to return empty-handed as usual.

After about half an hour a man came cycling up to us, and still panting, shouted: 'The fat Sahib has been killed (*maar dia*) by the elephant.' I was flummoxed. There was I sitting in the jeep, substantially larger than life despite my prolonged illness, and now this wild talk of the elephant having killed *the* 'fat Sahib'! What fat Sahib? We rushed to the spot described by the messenger to sort things out for ourselves. Obviously something serious and unscheduled had happened, but to whom? On our way we met a completely distraught Gopal Tanti, the department's most experienced specialist in chemical immobilization, brought all the way from Sundarban tiger reserve specially for this operation. He was wandering by himself, eyes wide, quite dazed—clearly suffering from shock. He said, rather incoherently, that the rogue chakna had charged the team and killed Vyas Sahib. He had no idea of what had happened to the rest. He was obviously in no state to give us a systematic account. Without wasting another minute we rushed to the spot indicated. When we reached the place we found nobody from the team; a few jawans hovered around. They told us that the army medical authorities, contacted over field telephone from the nearby watchtower, had already removed Vyas to the Corps hospital, and the others had followed in the departmental vehicle.

I had some difficulty in piecing together the whole story. It was a perfect *Rasomon*-like situation. There were several contradictory versions of what had actually happened. Here is one of them.

Apparently, the foot party had soon spotted the chakna, now by itself, in one of the quadrats of the supply depot that was not in regular use, overgrown with bamboo, bushes, cane, and interspersed with some trees. The chakna was heard feeding on a clump of bamboo about fifty metres from one of the camp paths that had a watchtower in one corner with field-telephone facilities—which is customary in such army depots or camps. In fact the watchers in the tower were the first to locate the elephant and pass the information to us. An open scrubby space separated the clump of bamboo from the road.

They approached the bamboo clump cautiously. The animal was chomping away happily on the other side of the clump. They had no heavy rifle, only a senior and a junior technician with one tranquillizing gun between them. It was well known to all that even if everything went according to plan, it would take at least ten to fifteen minutes for the drug to take effect on the darted animal. They could vaguely see the outline of the animal through the dense bamboo, but that was not good enough for a shot, particularly because of the possibility of the fired syringe being deflected by an intervening twig of bamboo in its path. There was a big log lying on the ground about 30 metres from the near edge of the bamboo clump, its top about four feet from the ground. The party climbed on to the log for a better view and, one suspects, the false sense of security that such a nominally elevated position offers the watchers.

Two forest watchers, brave boys, were carrying two departmental shotguns loaded with small bird shot in case of an emergency. There was a consultation in whispers on that moss-covered round log. Then somebody had a bright idea: if they talked loudly among themselves, the animal would be curious and come forward out of the bamboo, which would offer an opportunity to dart it in the open. As it so happened, the idea worked too well. The chakna burst out of the bamboo and charged full tilt at the party across the intervening narrow open space. All the members of the party standing precariously on the round log fell like ninepins off their perch, including Gopal Tanti carrying the dart gun. Vyas, unfortunately, instead of having a clean fall, slipped and fell slightly sideways *across* the log. So furious was the charge that the short-sighted animal did not even notice the fallen log on its way, and stumbling over it, fell head first on the other side, brushing over Vyas's prostrate body on the log. Vyas's guardian angel must have been working overtime that afternoon. While going over the log the belly of the animal slightly, ever so slightly, pressed Vyas's body. We saw later in the hospital that all of his body on one side was a huge contusion. Had the pressure been just fractionally greater, all the bones on that side of his body would have been crushed to pulp.

Then the animal tried to pick itself up from the ground. In the process, it threw back one of its hind legs and the horny sole brushed against one side of Vyas's face, taking off most of the skin. Had the leg been thrust out even a quarter of an inch more, his skull would have been crushed like an eggshell. Ultimately, apart from the state of complete shock he was in when we saw him in the Corps hospital, he was unharmed. The most thorough examination did not reveal a single broken bone in his body or any evidence of brain injury. This is one of the closest calls I have ever known.

We went into a huddle later in the afternoon. The problem was, if we abandoned the hunt at this stage, there was likely to be a loss of nerve. Officers and staff might not like to venture out on such an exercise again. So, it was decided that the exercise should go on.

D-day: 25 July 1988: The next day the morning routine was resumed as usual and at about nine in the morning, a message reached me at the Bengdubi forest rest house that the elephant had been darted and immobilized. I rushed to the spot at once. Here my past experience in Malaysia as an observer came in handy.

The local staff were used to Etorphine, a morphine derivative, which knocks out a darted animal completely. The usual practice is to administer the drug initially. The problem starts when reviving it with the antidote. Then with another drug, Rompun (*Xylazine hc.*), the revived animal has to be sedated, but *not immobilized*, and made to walk to the desired point or pushed on to a lorry with the help of koonkis. Our departmental experts had no experience of this kind of exercise. We had some footage on videotape of such exercises at Kattepura in Karnataka, and many slides taken in Malaysia, to give them confidence. That an unchained wild tusker could be made to walk tamely, tied between koonkis, was something new to the local departmental personnel.

Many technical problems had to be solved. We decided to follow the Northeast India practice of tying one koonki to each side of the captive to form what is known in mahout parlance as 'pakha' or wings, the captive animal itself being the body. This was also the system followed by the Malaysians, who, interestingly, had had their initial training from mahouts from Assam led by B.R. Phukan of Guwahati. We added a leg anchor following the Karnataka practice, that is, tied one of the hind legs of the captive to a koonki. If the koonki stopped and stood like a pillar on the command of the mahout with its four legs spread out for maximum purchase, in a position called 'khupi' in Northeast India, the captive animal would not be able to rush forward or charge, but would be pulled back and be drawn down on its hind legs.

At first the forest officers present insisted on having the front legs hobbled even while the animal was on the move led by koonkis. We realized soon that this too was unnecessary if one correctly understood the visible effects of the muscle-relaxing drug. It was more important to drag the captive animal as quickly as possible to the pre-arranged stall in the temporary camp, and tie it down for the night away from the madding crowd which always gathers on such occasions. In fact crowd control became a major problem while we were dragging the captive animal. Some koonkis had to deployed just to keep the road in front clear, and prevent the crowd at the back from closing in too much and pressing on the dragging team from the sides: a kind of job that the mounted police of Calcutta routinely carry out in the riot-prone football fields of the city.

Rajkumari was at first used as the back-leg anchor, but quickly proved an embarrassment in that role. So tender-hearted was she that at the first indication of distress from the captive, she would go *forward* to help the animal, thus creating a slack in the anchor rope and defeating the very purpose of the anchor. She was soon transferred to crowd control and replaced by Chandrachud, then a young departmental tusker of approximately the same height as the chakna. We felt by then that we could do without a back anchor. Rajkumari had not been much of an anchor, and that had not made a difference to the exercise. Meanwhile the chakna was showing the effects of a mild overdose of the drug: it was definitely becoming lethargic, and reluctant or unable to walk at a brisk pace. Chandrachud was asked to push the captive from behind with his forehead and bhusung (the base of the trunk as it is called in Northeast India) to cajole the captive to go faster. Unfortunately, we had not trimmed Chandrachud's pointed tusks, as he had not been originally intended for this duty. In fact, to be honest, we had not even thought of this as a possible function of a koonki. Chandrachud thought it great fun. Instead of pushing the captive with the front of his head, he started jabbing its back with his pointed tusks. There was no way we could restrain him. The backside of the captive soon developed an intricate design of little read poke marks, acquiring the look of an overused pincushion.

Then there was trouble from Jatra Prasad, the massive departmental tusker, who was being used as one of the 'wings' to lead the captive. Now, Jatra's past history was that he had, over the years, and from time to time, been gored by almost all the wild tuskers in the Dooars. I had always wondered why Jatra Prasad was targeted for these attacks, not other large tuskers in the pilkhana. The explanation came from Dinabandhu, a departmental mahout. Jatra had the 'bad habit', said he, of showing his huge, thick tusks to the wild visitors to the pilkhana. Now, how does a tusker 'show' its tusks when they are already visible enough? Dinabandhu explained with gestures that Jatra used to face a wild tusker, thrusting his own head and tusks skyward, no doubt with his ears suitably flared out, what in more sophisticated terms could be described as a 'threat gesture'. Naturally, the wild visitor felt honour bound to take up the challenge, and Jatra in chains had little chance against his adversary after having deliberately invited trouble.

The situation here was very different. Here was an animal, drugged and helpless, tied to his side. Jatra decided it was too good a chance to miss. To him it was an opportune moment to repay some of his past debts to his wild cousins. Intermittently he started taking sideways swipes at the captive with his thick tusks, ignoring the severe reprimands that the mahout administered with an iron elephant goad. The chakna was bleeding copiously from its cheeks and mouth by the time we reached the tying-up point in the temporary forest camp.

It was a particularly belligerent animal. The anxious forest officers insisted on repeating doses of Rompun several times on the way. As soon as the animal was tied securely to a tree in the camp for the night for transportation by lorry the next day, the animal just keeled over in a deep sleep, and was soon snoring gently. Even while in a state of drugged unconsciousness, its eating reflexes were unimpaired. Lying on its side, apparently sound asleep, it continued munching on titbits offered directly to its mouth. As we had no antidote for Rompun with us, we had to allow the chakna to sleep it off.

D-day + 1: The next morning by eleven we were ready to carry away the animal in a departmental lorry with a body modified for the purpose. We were to take the captive from Taipoo block of the Bagdogra forest range of the Kurseong forest division, where the captured animal was, in the extreme west of the North Bengal forests close to the border of Nepal, to the core area of the Buxa tiger reserve in the eastern part of the North Bengal forests close to the border of Assam, a distance of about 180 kilometres. The translocation was carried out under Rompun (*Xylazine hc.*) The time from start to finish was fifteen hours, including the time for weighing the lorry with the captive on it at a wayside weighbridge, traffic hold-ups, and several booster doses of Rompun on the way. We followed the arterial road connecting Assam and the Northeast with the rest of India. The heavy traffic on the road slowed down the progress of our convoy which consisted of the lorry carrying the captured animal, three more lorries with three koonkis for the unloading and release operations, and several jeeps carrying the officers and staff of the forest directorate. The captive animal had been sedated and loaded on the lorry with the help of two koonkis, a drill with which we and the koonkis were by now familiar after several days' rehearsal, using Chandrachud—then just in his early twenties and approximately of the same height as our quarry—as a substitute for the captive animal. Forest officers held an impromptu press conference in the camp while I, with some others, looked after the more mundane details such as tying up the captive animal on the lorry, loading the koonkis on their lorries, and so forth.

As mentioned already, we were working with a major handicap: we had no antidote for *Xylazine* with us to reverse the effect of the drug, if needed. We therefore decided to give a small dose initially, and give booster shots on the way as required. If the animal collapsed in the lorry due to an overdose of the drug, as it had done the previous afternoon after being brought to camp, we would be in serious trouble. It could easily collapse against one of the side flaps of the lorry, tilting its balance dangerously. The side flap of the lorry could even give way under the three-ton deadweight of the animal.

Several other important decisions had already been taken. The koonkis selected for unloading the chakna at Buxa were Jatra Prasad, Rajkumari and Urbashi. The convoy was to be preceded by a police pilot van using its siren to clear our way, which was necessary as we had chosen the shortest route to Buxa, avoiding the clearer but landslip-prone and twisting hilly stretch upstream of the river Tista. On level ground the highway however ran through the congested marketplaces of several small towns.

The captive animal had been loaded on the lorry facing the back. The Malaysian experience had shown that an animal coming even partially out of its sedated stupor, if facing the driver's cabin, could try to reach the front cabin by extending its trunk through a window, or could put its tusks through the back of the driver's cabin. Besides, it was easier to unload the animal if it was facing the back. We also tried to draw a lesson from the videotaped Karnataka experience which showed a huge killer tusker, sedation partly worn off, trying to scale one side of the lorry—at once a comical and a terrifying sight. Wise from this, we added an additional safety measure for which our model was, partly, Malaysia. Both the front and the hind legs of the captive animal were tied down to specially designed hooks on the floor of the lorry, which could be pulled up when required. These were firmly attached to the chassis below the vehicle.

Gopal Tanti, and Subrata Pal Choudhuri, then an apprentice to Gopal Babu, took their place on the roof of the driver's cabin, ready with darting equipment and drugs, to give booster doses to the animal when required. We followed closely behind, watching intently for signs of the animal's recovery, signalled by the animal trying to batter down the back flap with increasing vigour. By the time we reached the unloading point, the wooden planks of the back flap had been smashed to smithereens, and the flap was hanging by its iron rivets.

We wound our slow and weary way along the highway, heavy transport vehicles streaming past, up and down, in an unending flow. At around eleven at night, having come through most of the small towns and feeling somewhat relaxed, we stopped at a dhaba at Birpara at the crossing of the National Highway and Falakata Road for some overdue refreshment. None of us had had a bite or a sip since the morning; what was worse, the koonkis had not had anything but a few plantain stems thrown into the lorries at the start. It was well past midnight when our convoy left the highway and turned left at Cheko, beyond the Alipur Dooar crossing. The forest road would take us on the final lap of our journey to the appointed place with the specially designed ramp for unloading— or so we fondly hoped.

D-day + 2: We reached our destination, the heart of Buxa tiger reserve's core area at about two in the morning. A drizzle had started by then. Skidding and slithering along the forest road, very recently patch-repaired specially for the passage of our convoy, we finally reached the unloading point at the crossing of the 23rd Mile Road and Sangai Road, names of forest tracks as they appear in the forest map. At once we found disaster staring us in the face.

Instead of a U-shaped ramp, there was a massive, square platform sloping to one side. Only the slope followed the design, nothing else. The lorry carrying the captive animal was supposed to back into the hollow of the U. Then the two side flaps of the lorry were to have folded down flat on the two sides of the U, making platforms for the koonkis to stand on two sides of the captive. They would then have formed the 'wings' on two sides of the captive and pulled the animal out and down the sloping side of the ramp for its eventual release. The design of the ramp, therefore, was crucial to our purpose.

There must have been a lot of eerie beauty in the deep, mysterious forest around, but none of us had the eye for such trivia at that moment. To make matters bleaker, the local managers had forgotten to arrange for drinking water for the visiting party in that dry, bhabar tract; nor was there any food for us although we had been on the road, moving and partly moving, and sometimes not moving at all, for fifteen tense hours.

It was time to be firm. Before we discussed anything else, I said, first we must, abso-bloody-lutely must, have drinking water, hot tea, and something to munch. Fortunately, Rajabhatkhowa, the tiger reserve's headquarter, was only a few kilometres from that point, and luckily the unbridged forest river Bala, in between, was not in spate, as it had not rained the whole of yesterday, not even in the hills up in the north on the border of Bhutan. Two jeeps rushed back to Rajabhatkhowa and soon my demands were met.

My first reaction upon seeing the ramp was one of utter frustration. The carefully drawn design of the ramp had been ignored just as the natural requirements of the party travelling for the whole day and the better part of the previous night had not been thought of. I flopped down of the seat of the jeep in which I had travelled in the company of Mr G.S. Mandal, then Chief Wildlife Warden of West Bengal, and Mr S.C. De, Conservator of Forests (Wildlife). My mind went blank. There was a sinking feeling in the pit of my stomach which was probably not due only to lack of food and the nervous tension of the preceding hours.

It is surprising what cups of hot, oversweetened tea and some biscuits can do to clear the mental fog on a hard day's late night. I had earlier seen but not registered, two empty lorries nearby which had carried the boulders for the construction of the ramp. A post-tea mental flash suddenly revealed to me their possible use as the two arms of the U. Meanwhile the captive animal was still in its lorry, chained and roped. The koonkis had been let out of their lorries and were grazing on the tall grass fringing the forest roads. It was all peaceful and quiet in the sombre gloom dotted by pools of light from pressure lamps. All the turbulence and tension was in our minds: behind everything loomed the grey shape of the killer elephant in the lorry and our awareness that we just did not know how to unload and release it. People have millstones round their neck, we had this elephant. Maybe our escape route lay through those empty lorries.

The lorries were the same height as our captive-carrying lorry. The idea was to park them on two sides of the captive-carrying lorry, so that the koonkis could stand on them as they would on the two arms of an earthen U-shaped ramp. The captive animal then would be tied between them, and the koonkis would pull the captive animal out and down the ramp's slope for release. The trouble was that these just-appropriated lorries had rigid side flaps not specially designed to open out flat sideways. Some enterprising drivers and handymen from our own lorries, once they understood the problem, set about dismantling the side flaps. The operation 'trimming the lorry' was over in about 45 minutes. Both Rajkumari and Urbashi, however, refused to step on the platforms created by the part-dismantled bodies of the lorries which creaked, squeaked, and swayed under their weight. They were well-brought-up ladies of great dignity. Not for them such tomboyish pranks. No amount of cajoling from the mahouts could persuade them to be indecorous. Then we called upon Jatra Prasad as our last resort as he had been in numerous such past crises. He went up the sloping side of the ramp, smartly turned around under the mahout's command, and without hesitation, when told, stepped backward on to one of the de-flapped lorries. We did not exactly shout 'hurrah', but the collective sigh of relief of about thirty people seemed to go through the forest like a rustling breeze. The chakna received a heavy shot of *Xylazine* now, preparatory to taking off its chains and ropes, a tricky business which was accomplished by men taking advantage of the protection offered by Jatra's bulk. What a monument of cool assurance Jatra could be at such moments! It took nearly an hour to take off all the chakna's chains and ropes. Jatra then pulled the captive forward to the *terra firma* of the platform by the noose round the chakna's neck, tied to Jatra's girth belt (pharda). Urbashi next came up the slope from the other side to put the second noose round the captive's neck. The drugged animal was showing signs of recovery. The trunk now was not totally limp as earlier; the penis, extended fully under the effect of the muscle-relaxing drug, was already halfway back within its sheath; the eyelids were not as droopy as before, and the ears had started flapping. Time was running out. The mahouts were told to cut the two rope noose round the neck of the captive, and return double quick to the protection of the structure of the unloading platform. The fully revived animal could very well charge the captors. Our jeeps and lorries turned around, the engines started purring: we were all ready to flee at the first sign of a charge from the chakna, which was now standing by itself on the edge of the forest, fifty safe metres beyond the sloping end of the ramp. The sky was lightening by the minute. The dark outlines of the giant gokul (*Ailanthus grandis*) and champa (*Michaelia champaka*) trees were slowly emerging from the background, which was getting lighter.

We waited tensely in our vehicles for the first move from the chakna. A jungle fowl crowed nearby greeting the dawn of another jungle day. The spell was broken by a shot from Mr De's gun. He had decided to expedite the proceedings by directing a charge of bird shot at the chakna's rump. The animal slowly moved forward and melted into the shadows of the forest beyond. Our session with the chakna was over.

EPILOGUE

In about three weeks' time there were bitter complaints from the camp commander of Bengdubi, 180 kilometres away as the crow flies from where we had seen the chakna off, that the killer had come back and actually chased a jawan. Mr S. Dhandiyal, who was the Divisional Forest Officer in charge of wildlife in the Bengdubi region, penned a sharp reply, accusing the complainant of prevarication, warped imagination, and so forth, just stopping short of using the unparliamentary word 'liar'. He patiently pointed out that the animal had been captured and subsequently released in faraway Buxa. This temporarily silenced the army officer. But more such disturbing details began trickling in. Eventually one day the animal was found lying dead in the forest, a noose of rope still round its neck. By its death the animal proved that it had been alive and back.

The episode proved once again the homing instinct of elephants. An animal under heavy sedation had been translocated in a closed lorry over a distance of 180 kilometres but managed to return to its original point of capture in three weeks' time. This also demonstrated the inadequacy of the translocation of single animals as a management tool.

SECTION THREE

Managing Elephants in the Wild

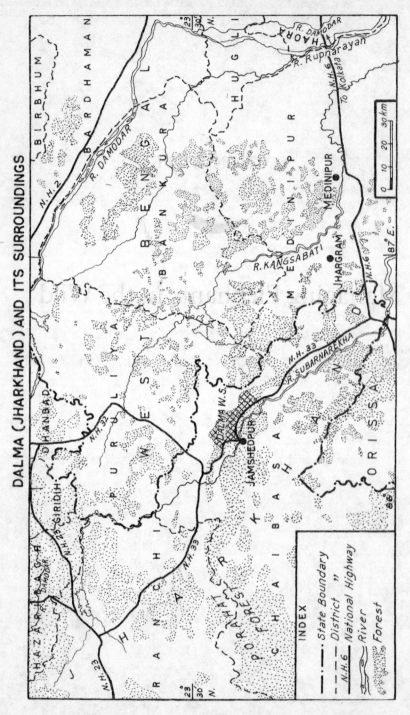

Dalma and its Surroundings

16

Introducing Dalma

I can claim an old family connection with the Dalma area. Before the reorganization of the states of India under Nehru, before the government took over the private forests in Bengal, Dalma used to be the property of the Medinipur Zamindary Company. In the 1940s two of my relatives bought plots there from the Medinipur Company. One of them even built a small shooting box there, and elephants used to come to drink water from the well in its grounds.

Dalma (c. 200 sq. km) is in the Dhalbhum area of what is now Singbhum district in Jharkhand. It is roughly to the west of the Purulia, Medinipur and Bankura districts of West Bengal. Extending east to west from Dalma are the Dhalbhum, South Chaibassa, North Chaibassa, Porahat, Kalhan, and Saranda forest divisions of Jharkhand. South of Saranda are the forests of Orissa: Bonai, Keonjar, and others. Then the range, after stretching down south, curls back east ending in Similipal forests in Mayurbhanj district in Orissa. The Similipal forests in their turn are connected with the west Medinipur, Ruam and Musabani forests of Dhalbhum, thus forming a loop, within the hollow of which nestle rural human settlements alongside giant industrial complexes like Tatanagar (Jamshedpur) and the Jadugoda mines, besides many smaller mining areas. This habitat is again linked with the Angul forests in Orissa, south of Keonjar. Dalma, thus, should be seen not in isolation but as the eastern fringe of the vast Singbhum–north Orissa range of elephants.

I came to know the elephants of Dalma in the mid-1970s. My friend Ashoke Kumar, then an executive of TISCO, and I were on the Eastern Region Committee of WWF-India. On his personal initiative and deft lobbying Dalma was notified as a wildlife sanctuary in 1977.

At that time, I accompanied Ashoke to Dalma at least five or six times a year. I was an active participant in the first few elephant censuses in Dalma, and was chased by elephants on more than one occasion, something that infallibly helps to establish one's credentials as an elephant researcher. In 1976, elephants migrating out of Dalma strayed out of their usual autumn home range—West Medinipur and Bankura—and entered Purulia as a result of a misdirected chase by the villagers in Bankura. The consequences were total confusion, disruption of civic life, and quite a few human deaths. About forty elephants were regularly sheltering in daytime under village trees, and watering in village ponds. At last the villagers and forest staff managed to push them back to Dalma by Patamda Road, a route the elephants had never used before and have not since.

In October 1987, elephants from the Matgoda forest range of Bankura crossed the river Kangsabati and entered the East Medinipur forest division, where nobody had seen elephants for at least the last century. One elephant calf fell into an unprotected village well. It was rescued and temporarily stalled in the grounds of the Divisional Forest Officer's bungalow. The officer's milkman was asked to supply milk daily for the young calf, and later the Divisional Forest Officer's informed us with some amusement that the milkman had never supplied milk of such high quality ever before. The milkman candidly admitted that it would be the greatest impiety supplying watered milk for the godly being that the elephant calf was, Ganesh Baba himself. In the milkman's mind, the Divisional Forest Officer was clearly not in the same divine league as the calf.

It was normal in the paddy season for elephants to move from Dalma to West Medinipur, and the Ranibandh and Matgoda forest ranges in Bankura in southwest Bengal. Now, having crossed the unprecedented rubicon of the Kangsabati and entered East Medinipur, they have been extending their autumn range every year further east and northeast of Medinipur. By 2003–4 they reached Durgapur, an industrial town in Bardhaman district.

The incursion of Dalma elephants into East Medinipur was not an isolated event. At about the same time, in 1986–7, a herd of about twenty elephants strayed out of the Porahat forest division and moved north through the Ranchi forest division and Gumla, and entered the Sarguja and Raygarh forests of Madhya Padesh. Within a few days twenty-seven people were killed. A huge controversy raged over whether these elephants should be allowed to continue in Madhya Pradesh. The Madhya Pradesh forest department was keen to have them as elephants had reappeared in the state after a long time. Bihar was equally vehement about not allowing the elephants to pass through it on their way to Sarguja, as the animals left death and devastation in their wake. Meanwhile, these elephants continued to-ing and fro-ing between Porahat and Sarguja. In 1986–7 thirteen elephants including a young tusker with strong man-killing propensities which appeared to have chosen Sarguja as their new home killed forty-one persons in Bihar and Madhya Pradesh. At last a hard decision was taken: these elephants could not continue in Madhya Pradesh. Consequent to this the Madhya Pradesh forest department launched 'Operation Jumbo' to capture these elephants. I was present as an observer during the first phase of the operation, which saw three adult animals captured, and subsequently the remaining members of the herd. Thus the Madhya Pradesh chapter of the story came to a close.

That year, another group of elephants moved out of Porahat to the Chauparan forests of Giridih–Kodarma on the edge of the Chhotanagpur plateau. These forests, associated indelibly with the memory of Bibhuti Bhusan Bandyopadhyaya, have everything—tiger, sambar, spotted deer and the rest—but no one had seen elephants there before. They went as far as the Pareshnath Hills across the Grand Trunk Road where, again, elephants were previously unknown.

It is difficult to explain fully this eastward thrust of the Dalma elephants in expanding their home range. This tendency to knock on the eastern door took a startling turn, again around 1986–7, when one small tusker, an adventurous soul, took the Hooghly–Bardhaman route, crossed the Ganga, and passing through Halishahar, an industrial suburb of Calcutta, marched east towards Bangladesh. One villager was so desperate to acquire a piece of ivory for himself that he confronted the young tusker passing peacefully through his village at night with a massive wooden club in hand, determined to break off a piece of ivory from the tusker. However, the plan did not go down well with the owner of the tusks and one more human casualty was added to official records. After a few more such incidents, which went to the debit side of the tusker's ledger, it was captured not too far from the Bangladesh border, and taken to Hollong. Unfortunately, it did not survive there long.

Wildlifers have searched for reasons for these synchronized migrations, even attributing them to the El Nino. But it is possible that the indiscriminate destruction of forests in Singbhum, particularly in Porahat, as well as gregarious flowering of bamboo in the region, induced these elephants to go out in search of a less disturbed home. The question, however, remains: why did these animals not explore the possibility of finding an alternative, relatively more suitable habitat further west or probe the excellent cover of Similipal in the south, certainly a habitat easier of access from Porahat than Kodarma.

In the past elephants had an extensive home range in southwest Bengal. They crowded Bankura; Purulia was entirely elephant forest up to the Rajmahal Hill through the Santhal Parganas. Pilgrims had difficulty in reaching the shrine at Deoghar because of elephants on the way. Perhaps their return to lost ground means a rehabilitation/restoration of the habitat, the elephant performing its role of an indicator species.

~

17

Return to Sender
East Medinipur, 1987–1988

I

When the elephants of Dalma crossed over unknown ground, as we saw in the previous chapter, the forest department wanted to push them back to their old home range, west of the river Kangsabati. Two departmental koonkis from North Bengal and later three hired koonkis were drafted into the operation. The man in charge of field operations, Mr S. Dhandiyal, was Divisional Forest Officer in charge of Wildlife Division I in North Bengal, and had considerable experience tackling marauding wild elephants.

This time, energized fences, a tool with which the forest department staff in South Bengal were not yet quite familiar, were to be used to restrict the movement of the herd to the western fringe of the forest division. The earliest such fence in southwest Bengal had been personally put up by the top people in forest administration: Mr G.S. Mandal, the Chief Wildlife Warden of the state, pulling the wire at one end and Mr S.C. De, then Conservator of Forests, Wildlife Circle, pulling it at the other, and Subrata Pal Choudhuri, the technical officer, putting up insulators, checking the tension of the wire and so on. The staff in North Bengal had known such fences since 1980, but this late exposure in South Bengal to the mysteries

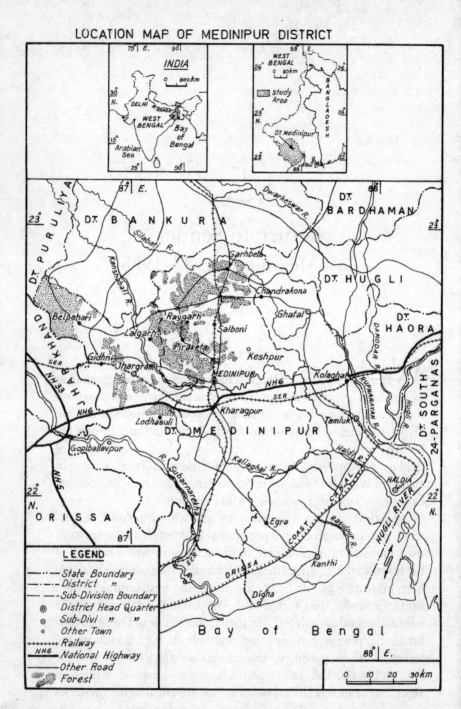

LOCATION MAP OF MEDINIPUR DISTRICT

Map of East and West Medinipur, West Bengal

of modern technology was no dampener to the remarkable innovativeness of the local staff here. In the initial days there were the usual hiccups. We realized only when actually putting up the first fence that we had no voltmeter to test the strength of electrical pulses flowing through the wire; nor did we have a line-testing light. But such small matters cannot hold back a forest ranger in full cry. Soon it was confidently asserted if we touched the wire with a green leaf, we would surely feel a tingling sensation if the line was alive. The next problem was trickier: what about the strength of the pulse coming through the wire? Surely, at this stage a voltmeter would be indispensable! No, not at all! Why not try the goat test? Just pick up a village goat and throw it from a short distance at the live fence and watch the goat's reaction.

In short, the first energized fence in southwest Bengal was put up by people with little technical knowledge of the instrument they were supposed to use.

To return to the herd that had strayed in from Dalma, the forest department staff led by Mr G.S. Mandal and Mr S.C. De located it one day in a patch of forest in the East Medinipur forest division. Using two departmental koonkis, they decided to chase the herd back across the unmetalled Salbani–Goaltore road, to the larger chunk of forest beyond. The chase started at Sijua in the afternoon. The police stopped the movement of vehicles and men along the road adjoining the forest to allow elephants to pass across the track without hindrance. There were more than a hundred scooters at each end of the blocked stretch of the road, gathered there to enjoy an afternoon's fun. The police requested them to leave the area, but the dauntless spirit of Medinipur—perhaps the most politically conscious district of West Bengal—refused to oblige. They claimed their inalienable fundamental right to gather on a public road to watch what was obviously regarded as free public entertainment. The police were helpless.

The party of two koonkis, Mr Mandal on one and Mr De on the other, along with some local staff of the department on foot, nevertheless started the chase. Things were moving according to plan and so, most obligingly, were the wild elephants. Just as the head of the first wild elephant emerged from the forest fringing the track, two hundred or more men kicked as one man at the starting pedals of their machines. The deafening noise of so many scooters suddenly roaring to life turned the elephants back through the line of beaters and koonkis. Things were back to square one. However, the same night they left the patch of forest, crossed the track, and went back to the main block of the forests of Abhaya to the south of the track.

The next morning, the koonkis were shifted to the grounds of the Ranja Beat Office, further west, as a part of the fluid strategy of following the herd, and keeping up the pressure. As one of the departmental koonkis was a young tusker, a two-strand energized fence was put around the koonkis' stall to prevent wild bulls accompanying the herd from attacking the tusker. Subsequent events justified the caution. Mr De and Mr Pal Choudhuri met a wild bull on the road in front of the Beat Office a few days later. The bull, no doubt irritable at being pushed around for days, decided to have a showdown with the intruders, and charged the party. A timely slug at close quarters from the departmental .315 bore rifle proved to be an effective deterrent, but it had nevertheless been a close call.

II

In January 1987–8, the forest department made another attempt to push the errant herd back across the river Kangsabati. The herd was reported to be in the forests of Bhadui beat of the Katapahari area under the Lalgarh forest range. Forest staff were instructed to watch the elephants and report their hourly movement—not a terribly difficult thing to do considering the patchy nature of the forest. By that time, in addition to the departmental koonkis, Ms Parvati Barua, whose koonkis were renowned, had also loaned three elephants to the forest department.

Wise from experience in North Bengal, we decided to start the chase late in the afternoon, preferably after 2 p.m., so that the elephants' natural tendency to move again after their mid-afternoon's rest would merge with the movement induced by the chasing party. There was a young tusker among Ms Barua's koonkis, recently purchased from Sonepur, which would have a crucial role in the operation to follow. The mahouts were instructed to march their elephants in the morning to a pre-arranged spot near the road on the fringe of the forest, and the local forest staff were instructed to intensify their vigil and keep a constant watch on the movement of the elephants.

We arrived at the prearranged spot at about two o'clock in the afternoon. The koonkis were there all right, but there was no sign of the forest department staff—they had, apparently, not returned yet from their midday tiffin. We bided our time—after all an army cannot march on an empty stomach. Eventually they started drifting back in ones and two, looking suitably impressed by the presence of the top nibs of the forest department. Unfortunately, when we enquired about the whereabouts of the herd, they looked at us bewildered. Apparently they had understood their instructions to mean watching the road and not the elephants sheltering in the nearby jungle.

We took the only available course of action open to us: we got on the koonkis and, accompanied by the local staff and villagers, proceeded to find out for ourselves where the elephants actually were at that moment. It took about forty-five minutes to sort out the details and get started. We proceeded along the dry, harvested paddy fields and semi-dry, narrow, watercourses fringing the forest which extended east to Jhitka and beyond, enquiring of the villagers returning from their daily chores if they knew where the elephants were. Nobody seemed to know. After about half an hour we met an old man with one crippled leg, returning from the fields. We asked him almost without expectation of reply if he knew where the herd of elephants was. But he had seen the herd entering a patch of sal coppice forest some time earlier, and agreed to show us where. We were an incongruous sight: a small army of five elephants and a large number of 'foot-soldiers', equipped with firearms, crackers, and rockets led by the lame man hobbling on his stick. As we approached closer, sure enough there were fresh marks of elephant tracks on the loose soil around the edge of the forest.

After a few minutes we reached the edge of the forest where the old man pointed his stick. Almost immediately we heard the rumbling of elephants beyond the green curtain of sal. Bidding the old man goodbye, we entered the forest in an Indian file, one person behind the other, the elephants spaced out between the men on foot. We had clear instructions. At the sound of a whistle, men and elephants would wheel around 45° on their axis and transform themselves into a 'shikari line' moving forward shoulder to shoulder, like a line of beaters. 'Chance' had no chance in such a meticulously planned exercise. It fell to my lot to ride on Ms Parvati Barua's new tusker along with Subrata Pal Choudhuri. Mr Mandal was on a departmental tusker. Mr Dhandiyal like a true leader of men, decided to stay with the men on the ground.

After we had proceeded for about 50 metres in the forest, we caught up with the wild elephants. A young tusker about 25 metres away was dusting itself under a mahua tree, glowing like tarnished gold under the beams of the setting sun. The mahouts were carrying, as instructed, some loud firecrackers. Keeping the .22 rifle I was carrying menacingly ready to meet any emergency, I instructed the mahout to lob a cracker at the elephant enjoying its dust bath. With the loud explosion, everything went still for a moment. Then the whole jungle exploded with what sounded like a vast echo of the blast we had made as more than fifty elephants went trumpeting and crashing through the thick, young coppice growth. We had to keep up the chase and crash through the forest ourselves behind the receding herd.

At this stage all our well-laid plans went awry. A hunting line is all right when the line is moving sedately through the forest scouring the cover systematically for game; but chasing a fleeing herd of elephants through thick cover, we realized rather late, was quite another matter. In such cover, elephants can move much faster than men, who are not equipped physically to crash through obstacles on their way, and domesticated elephants are no match in this game for wild elephants unencumbered by such impedimenta as mahout, riding gear and riders. In the heat of the chase, the koonkis got separated from each other as well as from the men on foot, each koonki going its own separate way following the nearest trail of sound. In the pre-mobile, pre-walkie-talkie days, soon all communication was lost between the different units of the chasing party. In the short winter evening, at one customary stride came the dark.

After blundering around in the dark for some time in dense cover, I realized we were totally lost. None of us were equipped to read the stars for directions, but one thing forest guards seldom lack is confidence. The forest guard with me and Subrata confidently pointed out the direction of the nearest Forest Beat Office, Bhadui: first, he pointed straight towards the west; after that yielded no Beat Office, he pointed northwest, then southwest, and once even eastward. It went on like this for some time. Sometimes we could hear the crashing of wild elephants nearby. We kept calling out towards the Beat Office which was alleged by the forest guard to be nearby, hoping that all this shouting would keep at least the wild elephants away. Blundering in the dark through dense cover with more than fifty wild elephants milling around, we had no way of telling how close we were to them, as sound, our only lead, was not a very reliable guide in these circumstances. (It is remarkable the distance the sound of an elephant's rumbling can carry in a still night. The same thing has been observed of the lion's grunts.)

Meanwhile, an uneasy feeling was growing in me that our party was guaranteed to get unwanted company. We were especially vulnerable as our mount was a young tusker, prone to attack by a wild tusker. Interminable dark minutes away we glimpsed a patch of light to our left, fleetingly appearing between poles of sal coppice, somewhat high up. Assuming we were marching straight west, that would be in the direction of the Bhadui Beat Office. In a few seconds, we suffered some pangs of doubt: could that be the light of some village? After another half an hour or so of brushing through the undergrowth and dodging the dark sal poles we emerged in the open; there was the assurance beyond in the open fields surrounding the Beat staff quarters. A thoughtful pressure lamp was burning from a tall pole, acting as a beacon.

My nagging fear of a wild tusker tracking us, reversing the roles of the hunter and the hunted, was now realized. On reaching the open ground, hearing a twig snap somewhere behind, I turned my torch in that direction, and sure enough a wild tusker was just emerging from the forest edging the field, about fifty metres behind us. Much more limber than I, Subrata had already slid off our mount to the ground. In panic, I shouted at him to take a rocket from his backpack and fire it towards the elephant, the mahout counter-manding this strategy with an equally panicky, 'No! No!' I remem-bered too late Parvati's warning that her recently acquired koonki had not yet been broken to the sharp hiss-and-bang of rockets. As the rocket went off, our wretched mount wheeled round on its axis and started shaking its body to rid itself of us, very much in the manner of a wet dog shaking water off its coat. We hung on des-perately to the harness ropes trying to cling to our seat on our precarious perch, the sudden looming of the wild tusker from behind adding urgency to our effort. When things settled down a bit, relatively speaking, another sweep of the torch beam revealed that we were rid of our unwanted company. Slowly, one by one, the men and elephants drifted out of the forest guided by the hanging lamp and joined us. Thus ended the day's misadventure, teaching us some drawbacks of driving an elephant herd at night.

III

A third attempt was soon made to drive the elephant herd away. It was located, then, between Katapahari and Jhitka. It was to be a night drive to push the elephants to the Kangsabati and beyond, to the original home range of the herd. Mr S. Dhandiyal was again in charge of the drive. Koonkis were not used in this drive, because of the risk to the koonkis from wild males in such a night operation.

The hoola party (party with lighted brands to chase and frighten away elephants) started the drive at about eight in the evening. Responding to the drive, the herd was moving in a westerly direction towards the river Kangsabati. To repeat, our hypothesis was that these elephants were lost in a territory unknown to them, and all we needed to do was to lead them back to their old stamping ground, and they would then move away on their own finding their bearings. There were about three or four powerful spotlights in the driving party, each light attached to a car battery, slung on a pole and carried by two persons to ensure mobility. For those unfamiliar with the use of spotlights as a deterrent to elephants, elephants as well as leopards and pigs are not usually frozen by strong beams of light as tigers or deer are, but tend to shy away from the light. This has been a standard piece of equipment with anti-elephant depredation squads in North Bengal since at least 1976. In South Bengal the practice was only a recent replication of the North Bengal one. However, a beam of light that is not strong enough, as from a hand-held electric torch, is liable to provoke rather than deter a charge from an elephant, as these marauding elephants seem to associate such lights with bodily pain inflicted by villagers when guarding their crop in the field.

The specially recruited hoola party began the chase. Soon they were joined by lustily shouting villagers holding up lighted brands. The spotlights were spaced out between them, strengthening the chasing line. Very soon Mr Dhandiyal and I realized that we had no hope of keeping up with the chasing line on foot over the broken ground, a line that was fast moving away from us. We then decided to call for the jeep parked somewhere on the road between Lalgarh and Ramgarh and then intercept and join the party on the west bank of the Kangsabati.

By the time we reached the Kangsabati in our jeep, both the elephant herd and chasing party had already crossed the river, which at the end of the winter, was running about knee-deep. All the electric lights we had, spotlights as well as hand-held torches, were with the chasing party. I had one pen-torch and my trusted .22 bore rifle with me to face any crisis. On the other bank of the river we could see suddenly spurting spotlight scarring the night sky, some trees and bushes fleetingly lit up by the lights, and hear shouts of people, occasional crackers and squeals of elephants. Then our group and the elephants moved further away. We waited on a sandbank near the west bank of the river for the men to return. We estimated that their task was probably accomplished by now. The elephants we were chasing were, by our guess, now looking around themselves on the other side of the Kangsabati, and realizing they were back where they belonged.

As we waited, expecting our people to come back across the river, we heard a sudden continuous splashing of water, as if we were near some rapids or a waterfall. But how could a waterfall have developed suddenly in the middle of the flat and sandy Kangsabati which was so easily fordable at that time of the year? We were even more intrigued when the sounding cataract—the forest guard had calmly pronounced the sound to be a waterfall's—suddenly was sounding no more nor haunting us like a passion.

Eventually our party began returning in small groups, looking and sounding slightly sheepish. They had lost the herd of elephants after crossing the river. Now, how does one 'lose' more than fifty elephants in a reasonably open country? Well, as our men did, and as anybody else could have done under the circumstances. If one is using spotlights in an area with trees, cottages, and clusters of bamboo, beams of spotlights and electric torches are reflected back from these obstacles and what one sees beyond are fanning-out patches of shadow. In that cone of shadow even a whole large herd of elephants would be invisible to men near the lights.

We still assumed that the elephants had gone further west, as we had intended them to go. But could they have doubled back? As Mr Dhandiyal was coming back towards me and I was going forward to meet him across the sandbar, we saw elephant tracks on the wet sand between us. The tracks suggested only one elephant, or at most two. One or two coming back would not make any difference if the herd were moving in the westerly direction, driven by the chase. The villagers and the hoola party went back to their villages and camps in the departmental mobile vans, and we went back to our jeep. After a council of war with the Range Forest Officer, Lalgarh, we decided that we should keep vigil on the road for the rest of the night to prevent the elephants from coming back. All night we kept vigil, shuttling in two jeeps on the Lalgarh–Dharampur road which the elephants had to cross to re-enter the main forests beyond at Katapahari. When tinges of orange lightened the eastern sky, we felt happy that we had done our job well. There was no sign on the road of elephants having crossed it back. The first teashop opened in the marketplace. We persuaded the stall owner to give our party some tea. It had become quite chilly early in the morning. Now ground mist was forming around us and a low haze was hanging on the ground. As we sipped our tea we felt a slow-growing warmth of satisfaction of a battle fought hard and won.

Soon enough we had the company of a gentleman from a nearby village who also had had an urge for an early-morning cuppa. Even as we were preening ourselves on our successful night vigil, the gentleman began muttering imprecations: the herd had returned. It had demolished his standing crop, a snack in passing, before crossing the road from the west to the Katapahari forests in the east.

A useful moral was driven home to us: the elephants were not lost souls in an unknown land. They knew exactly what they were doing and where they were. They came because they wanted to come and did not take the way back to West Medinipur and Purulia because they were not ready to do so yet and our coaxing had not been strong enough.

18

Vet in the Forest
East Medinipur, 1991

In April 1991, the East Medinipur forest division sought the help of the Wildlife Wing of the Forest Directorate of West Bengal. A young tusker accompanying a herd of more than fifty elephants was injured and could not walk well; attempts to drive away the herd were failing, as the herd invariably came back after being driven away two or three kilometres, to rescue the injured teammate lagging behind.

We arrived at the Bishnupur forest rest house—S.C. De, then Chief Wildlife Warden of the state, Subrata Pal Choudhuri, technical officer specializing in tranquillization, and, I, a hanger on. It was getting uncomfortably hot, though it was only the middle of March. The herd and the injured elephant were constantly moving between different forest ranges; hence the local forest staff could give us no precise information about its immediate location. We decided on the next available option: asking the villagers themselves in the interior. On 17 March, we met a young tribal man on a village road who knew where the injured elephant was. As the animal was unable to move much, he said, he was taking it cut branches of trees daily to feed it. We were enormously impressed. We were in the heart of an area that was regularly ravaged by wild elephants during the crop season. The worst sufferers were the tribal people with their small landholdings; yet the young man was looking after the injured elephant. He gave us more or less precise information about the location of the animal. We came back to camp, and decided to pursue this latest intelligence report the next morning.

The place was called Dhabani. It was somewhere west of Garbeta and Mandal Pushkarini. Our team was now joined by the veterinary officer of Goaltore, who had no previous experience whatsoever of treating injured elephants, domesticated or wild. Villagers' reports said that the elephant had an injury in one of the front legs, which had swollen to double its normal size. Taking even one step was an agony for the animal.

The next day our jeep bumped across forest tracks and fields to the spot, where we found a group of enthusiastic young men had already gathered. Yes, the injured elephant was there all right, they said, but it was in the middle of the herd. It would be impossible to isolate the animal from the herd and treat it. After some cogitation, we had a bright idea. If we first chased the herd away, the injured animal would surely lag behind, unable to keep up with the fast pace of the fleeing herd. Explaining our plan, we asked the assembled villagers if they would be able to chase the herd away from the spot. 'No problem' was their ready answer, and in a few minutes the chase started. We could hear the herd crashing through the forest, the chasing party shouting and bursting crackers till the sound died away in the distance. The sun hung over our heads by now. The heat was oppressive.

The next step of our plan of action began. We asked the young village sparks if they could fan out in the forest in ones and twos and locate the injured elephant left behind by the stampeding herd. After about half an hour or so, two boys came running and told us that they had found the elephant, standing by itself in one corner of the forest. There was a jeepable track through the forests to the spot. We set off for the spot in our jeep with our two young friends as guides. Another ten minutes, and we were there, but we could not immediately spot the elephant. The elephant, disturbed, had apparently moved away a short distance. A five-minute walk through the young sal coppice brought us to the tusker which was standing on a slight elevation just across a depression created by a dry nullah. It was at a distance of about fifty yards, offering a quartering-away view. Subrata was asked to prepare his syringe. His experience so far had been limited to tranquillizing mostly langurs, where time was not of the essence. He had therefore not thought of keeping prepared syringes ready. Showing an exemplary cool, he and his assistant Prasad spread a cloth on the ground and arranged all the phials on it neatly and systematically, taking all the time in the world. Standing a few feet behind him, we were urging him in hoarse whispers to hurry. We had to be careful not to hassle him too much. By that time a small crowd had gathered behind us ready with a battery of advice.

By the time Subrata was ready, the elephant, slightly disturbed, had begun moving, limping visibly. Too many sal poles intervened between us and the animal for an accurate hit in the right spot. With a .315 departmental rifle Mr De stood guarding the precious box of drugs and other tools of trade of the darting party. Subrata went forward with his darting gun, and I followed to give him moral support and advice. Very soon we caught up with the elephant slowly limping away in front of us. But as it was moving one could not be sure of hitting the right spot, which in this case was the fleshy part of the elephant's rump.

We moved forward slowly, only thin poles of sal between the pursuers and the pursued, Subrata always about five feet ahead of me. Suddenly the elephant stopped. I clutched Subrata by his shoulder and whispered into his ears, 'Fire'. Up came the dart gun. I could hear and see the dart hitting and sticking to the rump. The animal winced slightly, registering the hit, and then, pricked by the dart, it quickened its pace. We quickened ours accordingly, to keep up with it. In about five minutes its hind legs buckled under its body, and then it slowly keeled over on its side. A couple of hundred admiring spectators were standing about a hundred metres away. Mr De stood guard over them to prevent them from rushing forward.

We approached the animal and closely examined the eyes. There were faint signs of mild cyanosis in the membranes of its open eyes.

The breathing was regular. There appeared to be no immediate cause for worry. We called for the vet, standing with Mr De. He confirmed that the body temperature and the pulse rate were normal for an elephant. There was a huge traumatic injury around the left knee. The vet got to work. A deep incision across the swollen part of the leg brought out a flow of black blood. There was no other visible injury on the elephant and nothing inside the wound. The animal must have got injured stumbling over a cut piece of bamboo or sal stump. After the necessary treatment of its leg and a shot of long-acting antibiotic, we bandaged the leg—more for visual effect than as part of the treatment of its injury—we knew perfectly well that the bandage would not stay for a moment once the elephant was up on its legs after being revived with the antidote.

We knew the immobilizing chemicals would wear off any minute now, but the elephant showed no inclination to get out of its drug-induced sleep, which every well-bred elephant should after a few minutes of receiving a shot of the antidote. The ever-helpful villagers brought buckets of water from a nearby stream to pour on the drugged animal to keep its body temperature down. They also brought back the disturbing news that the herd was back and was making quite a bit of racket just beyond the stream.

The elephant, having made a couple of half-hearted attempts at lifting its head fell back into a deep sleep, gently snoring. We had no alternative but to wait for its revival and allow nature to take its course. As conscientious managers we should have stayed around for the moment of its revival. Unfortunately, there was not one tree strong enough to support a machan on which we could spend the night: it was all sal regrowth about two years old. Clearly, it would not be possible to spend the night on the ground with about fifty wild elephants a couple of hundred metres away, straining at the leash to come to the aid of their fallen comrade.

Reluctantly we returned to the jeep where a rude shock awaited us. All the drinking water we were carrying in the jeep in a large jerrycan had been drunk by the people standing around the vehicle. We had not bargained for this degree of people's participation. I had never felt so parchesd in my life. Subrata, predictably, played the ministering angel at that moment. He had one musambi in his pocket saved from his breakfast. He cut that little fruit into eight pieces, sharing them with us so we could moisten our tongues, a fine demonstration of *esprit de corps*.

In retrospect, perhaps the mixture of Etorphine (a morphine derivative) and Acepromazine maleate (a tranquillizer), blended in the forest, had inadvertently tilted more in favour of the latter. In those days, the forest department did not use readymade mixtures of Etorphine and Acepromazine, such as Immobilon. Trying to mix potent drugs under conditions of extreme stress has its pitfalls, especially with a wild elephant a few feet away. The elephant, we later found, had revived during the night and the herd had come and collected it.

Subrata's subsequent successes in tranquillization include sedating an elephant hit by a railway train just enough to leave him sedated but standing, for easier treatment. He also came to the rescue of a foolish elephant that had put its foot through a tyre in the Bengdubi army depot in North Bengal. The tyre had to be hacksawed out with the elephant still standing in it. He once tranquillized and brought down a tiger in the Sundarbans which had taken daytime refuge atop a date-palm, about thirty feet tall—the only recorded instance I know of a tiger climbing a straight-boled tree that high. In 2002, when trying to tranquillize a rogue elephant, Subrata was nearly trampled to death. He is not fully fit yet, perhaps never will be again, but his spirit remains indomitable as ever. Till he was put out of action he had tranquillized 66 wild elephants.

POSTSCRIPT

The injured tusker was killed in an intraspecific fight by a larger tusker the next year. The same veterinarian from Goaltore was asked to inspect the carcass and report on the cause of death. He had no difficulty in recognizing his previous year's patient with knee injury. Natural selection had eliminated the less fit of the two.

19

Right of Way
Chinsurah, 1993

New Year's Eve and its preceding days are festive in a metropolitan city. I celebrated the night of 30 December 1993 and the wee hours of the 31st not too far from the sounds and lights of Calcutta's rollicking Park Street, but in ambience a world away.

On 28–29 December the Dalma herd moved out of the Bishnupur forests of Bankura to Hooghly district; moving north it then reached Dhanekhali, famous for its fine cotton saris. The herd took a south-easterly direction, apparently because it was chased away by people.

I had read about all this in the morning papers, but innocent of what the chain of events held for me, I went to attend a meeting at Writers' Building, the seat of the state government. Making the most of Calcutta's fleeting cold season and the formal occasion, I was decked out in my winter finery: fashionable check tweed, but not too thick, matching tie, thin-soled leather shoes. Immediately after I reached the chamber of Banamali Roy, then minister of state, in charge of forests, I was practically abducted by Prabir Sen Gupta, who was then power minister, and MLA from Chinsurah. Formerly a Dutch settlement on the western bank of the Hooghly, Chinsurah was a part of greater Calcutta, and it was here that the elephants had decided to spend New Year's Eve. Consumed with anxiety about his constituency, Prabir Babu wanted me to accompany him to Chinsurah right away, straight from Writers' Building.

I pleaded lack of proper clothes and shoes for the occasion. These irrelevancies were brushed aside by Prabir Babu with his characteristic panache. The next problem was getting hold of a senior forest officer to accompany the ministerial party. The only such officer immediately available in town was J.N. Bhadury, then a Chief Conservator of Forests. Since he had once done a stint as Conservator of Wildlife, he qualified, and his fate for the moment was sealed. A peremptory phone call from the minister's room asked for his immediate presence. Half an hour later, Mr Bhadury reached the minister's room and was told he would have to leave for Chinsurah immediately, accompanying the power minister. His jaw dropped. He made the usual excuses about pending files requiring immediate attention, commitments at home, not feeling too well and so on. He was no more successful with Prabir Babu than I had been. Prabir Babu's argument was that it was only a matter of a few hours. We would nip over to the site to give moral support to the staff and, pronto, nip back to cosy, festive Calcutta—it was as simple as that.

Asking for a map of the area, I was given an ancient small-scale revenue department map which did not even show the Grand Trunk Road (National Highway 2) bypass, let alone the alignment of the Durgapur Express Highway, then supposed to be under construction. I requested some office staff at Writers' Building to phone and inform my wife that I might be a bit late returning home.

Our convoy whizzed along the Delhi road led by the red-light-crested white ministerial Ambassador. At dusk we reached the crossing known as Sugandhar Mor and turned left for the Durgapur Express Highway. We halted by a roadside police outpost for information about the latest location of the elephants. The police radio crackled and they informed us that the local government staff and villagers were trying to chase the herd away. The local police said they were trying to drive the elephant herd back to Dhanekhali along the Durgapur Express Highway, and from there back to the forests of Bankura across the river Damodar.

It was dark by then. Proceeding southeast along the half-finished

Durgapur Express Highway, we soon came face to face with the herd. We could see shadowy forms of elephants silhouetted against a ring of lighted brands, moving towards us along the highway. On the other side of the elephants and the ring of light were the heads of the district administration and their men. The elephants were on the highway moving northwest chased by villagers with lighted brands. It was a grand spectacle: dark, bobbing, almost floating, forms of elephants, silhouetted against the red light of the waving brands. They were steadily coming towards our convoy of three vehicles. Clusters of people had flocked by the road, loath to miss out the parade. The elephants and the men behind them were rapidly closing in on us, and lacking radios, we had no way of contacting the men. It was now time to clear the road and move to one side. We parked our vehicles on high ground just off the road, facing the approaching wave of elephants and the waving brands. I confess I felt slightly uneasy standing next to our white Ambassadors which are *de rigueur* for government higher-ups. I should have felt much more at ease with dark-coloured, less conspicuous, plebeian Gypsies.

The approaching line of elephants stopped before a culvert, less than a hundred metres from us, and the wise animals decided that they did not like it at all—a piece of PWD construction. The coaxing from behind became louder; there were one or two shrill squeaks of alarm, the line wavered, retreated a few steps scattering the brand-waving villagers, and left the road for the dry fields to the east. The line of flaming brands also broke rank and poured into the fields after their quarry. Obviously, things had gone awry, and defying all the careful planning of the district authorities, the elephants were showing that they had a mind of their own.

Standing in the dark we could see the ring of light moving across the crop fields away from the Durgapur road. We were soon joined by the district officials who had been bringing up the rear of what had been almost a ceremonial cavalcade led by elephants and men clad in reddish light. I had already told Prabir Babu of my apprehensions: the elephants were trying to break out of their usual home range towards east. I had in mind especially the case of the young tusker of the same Dalma herd which a few years ago had crossed the Ganga at Halishahar not too far from Chinsurah and marched eastward towards Bangladesh. The evening's movements were for me portends of more adventures in that direction. Having lost all sense of direction in the darkness, I asked Prabir Babu if the herd was taking an easterly direction. My heart sank when he said yes. East was the direction of the Ganga. Across the Ganga was the main body of greater Calcutta. I had visions of the herd crossing the Ganga at night and after a brisk march sheltering the next day in Barrackpore Lat Bagan, which was formerly the weekend resort of the viceroys stationed in Calcutta.

We had little time to discuss our plans with the district officials. Meanwhile, watching the moving light of the villagers' torches the minister, who knew each village in the area, confirmed the herd's eastward inclinations. Where we stood, on the west bank of the Ganga, were the towns of greater Calcutta on its north-western edge: Chinsurah, Hooghly, Chandannagar, Srirampore, making for an uninterrupted urban sprawl stretching to Calcutta and Howrah proper; on the other, eastern, bank of the river, were the North 24-Parganas district, and towns like Naihati and Barrackpore. Once the elephants had crossed the Ganga, how were we going to make the herd recross the river and push them back through the urban conglomeration on the west bank to the forests of Bankura beyond?

By then Prabir Babu had grasped the crux of the problem. The herd's eastward movement, towards the Ganga to be precise, would have to be blocked. It was therefore useless to chase it as the villagers were doing. It had become clear that the herd was not being chased according to any plan. The elephants were moving as they pleased, and relays of villagers wielding flowing brands were just running *after* them hastening them on in the direction they had already chosen. The villagers' sole concern was to drive the herd off their own paddy fields. There was hardly any forest department staff assisting and directing their efforts, except for a few accompanying their boss in his jeep. The matter had exploded so suddenly that there had been no time to gather more experienced staff from the neighbouring forest divisions. Besides, no forest staff could have run after the herd continually over about thirty kilometres. The villagers were doing it in relays: as soon as the herd left the paddy fields of one village, the farmers of the next would take up the chase.

Clearly the herd would have to be *intercepted* to turn it round. Prabir Babu explained that there was a road at a place called Kadampur southeast of the Durgapur Express Highway and running parallel to it, where we could possibly attempt the interception. We immediately rushed to that spot to block the herd there. All that rushing, however, was of no avail. The villagers were unprepared for the sudden appearance of the herd. We did not have equipment such as powerful spotlights to turn the herd. Before we had even parked our vehicle, the entire herd crossed the road in front of us in a run.

Prabir Babu, now aware of the hair-raising problems in store, explained that the only other suitable place to intercept the herd was the Delhi road. On our ancient map, he marked for us a point near the Sugandha junction where the herd could try to cross the Delhi road.

We reached the Delhi road before the elephants, at about ten at night. We decided to block the road with lorries. J.N. Bhadury, renowned for keeping a cool head in any crisis, and never mixing up his priorities, suggested a quick meal of dal and roti at a nearby dhaba; the Conservator of Forests knew it would be a long night. I accepted his wise suggestion with alacrity. Meanwhile the Superintendent of Police had ordered his men to stop all vehicles passing through the Delhi road and make them turn to face southwest, the direction from which the elephants were coming. The first reaction of the travelling lorry drivers to obstruction by the police was one of utter indignation. Why, oh why? Haven't we paid our customary homage just a few miles back? Then why stop us again?

However, they entered the spirit of the game when it was explained that their mission was to block the road. In no time at all more than a hundred heavy transport vehicles had lined the road all facing west. We explained the drill to them. When they heard a whistle, they all were to switch on their headlights and press on their electric horns, in short, push the level of the *sonne* and the *lumiere* to a level unacceptable to any well-bred wild elephant. A sprinkling of Marutis and Ambassadors had joined the string of lorries by then, all sportingly agreeing to contribute their twinkles and peeps to the full-throated cacophony of the electric and air-horns of a hundred or more five-ton monsters of the highway.

Across the plain fields swayed the extended ring of torches silhouetting undulating dark shapes. The edges of the shapes acquired definition as they drew nearer. The undulating mass was resolved into a broken line of bobbing elephant heads. The shouts of men chasing the elephants were now clear across the open ground. The point was to choose the right moment to put on the sound-and-light show. I had a policeman next to me, ready with his police whistle. There were two other whistle-blowers placed along the line, waiting for the first signal. I was standing next to a five-tonner, which seemed to be carrying double its licensed capacity, ready for a smart bit of footwork to dodge behind the back of it in case the exasperated pachyderms decided to break through the line of lorries. I had earlier seen indignant elephants unshaping and capsizing impudent Ambassadors and cheeky Marutis. The fully loaded five-tonner looked safe enough—under the circumstances.

About a hundred metres from the road the string of elephants broke into its individual components. It was time to send them the message: this far and no further. I shouted to the policeman to blow his whistle. Other strategically placed whistlers chipped in. The field was flooded with the headlights of the lorries and their blaring horns, with some squeaks and twinkles from the light passenger cars.

The line of elephants halted and then mysteriously oozed out of the arena lit up by the headlights of lorries. The villagers running after the elephants started appearing on the highway. One of them blamed us for his failure to capture a baby elephant which he wanted badly as a pet and did not hesitate to complain bitterly to Prabir Babu, his minister and MLA. For some time, strange though it might sound, we just lost those fifty or so elephants that had been there, larger than life, minutes earlier on a stage awash with light. Only a sliver of a moon floated gently across the slightly misty blue of the sky. The men of the chasing party, as dazzled by the headlights as their quarry, also had no idea where the animals had gone.

We shot out in a police jeep to explore the Sugandha–Durgapur track just to make sure that the herd had not moved across it to the north. We met some locals on the way. They seemed unaware of any movement of elephants across the road. Lights were still burning in some roadside houses. No, they were not aware of any elephants either. We drove for about five miles towards the Durgapur Express Highway. There were no tell-tale signs like dung boluses and broken twigs on the road. This, obviously, was a blind lead. We charged back to the Sugandha crossing where we met the forest department jeep and the Divisional Forest Officer. There was a village road leading southwest from the main road. The villagers, had told the Divisional Forest Officer the elephants had taken shelter in a big patch of banana and cultivated bamboo on the southeast of the road. The minister had already gone to the spot indicated by the villagers. The Divisional Forest Officer, as protocol demanded, was waiting for the Chief Conservator of Forests, Mr Bhadury.

There was a small gathering of locals surrounding the minister standing outside his white Ambassador. The villagers confirmed that the elephants were indeed sheltering in the large patch of bamboo and banana, east of where we were. We discussed possible plans of action. If we charged the herd with crackers from the western edge of the cover where we were standing, there was every likelihood of the herd moving east across the Delhi road, this time ignoring the barrier of the lorries and breaking for Chinsurah town and the river Ganga on the town's eastern edge. This would be easy as some of the lorries had given the slip to the police by then, thus creating gaps in the mobile roadblock we had put up. If we pushed them from the south, the herd could move north parallel to the highway, entering the large village on the way and trying to pass through it. Both eventualities contained the distinct possibility of accidental human casualties. The third and the most sensible alternative would be to leave the elephants where they were, tighten the cordon on the village road and resume the operation in the morning with reinforcements of hoola parties from Bankura and Medinipur.

However, Prabir Babu, standing soaked in moonlight, shook his head firmly. Ministers are made of sterner staff. He had already decided that it was nearly two in the morning, the herd was to be driven out from their shelter at once without waiting for daylight, contrary to plain commonsense. The minister ordered the DFO to move into the cover with the few men he had at his disposal 'to do the needful'. The DFO, to the credit of his good sense, flatly refused to oblige.

We now had an unpredictable herd of elephants as well as a furious minister on our hands. With a grunt which almost rumbled through the sky, he marched straight ahead along the narrow village path flanked by his two security men. We watched in stupefied silence at the kind of valour Rana Pratap of Haldighati fame was renowned for: famously brave, but perhaps with a touch of foolhardiness. As Prabir Babu disappeared into the elephant-infested gloom ahead, with his two paragons of parade-ground rectitude, we could make out their forms in the faint moonlight: and suddenly there was a gap like a missing tooth in the posse of three; while Prabir Babu was briskly walking forward, his two guards were transfixed to the ground albeit in the correct martial position, legs planted slightly apart, shoulders square, chest out, stomach in. The suspense continued for a few minutes. Prabir Babu returned as briskly as he had advanced. His escorts made a smart about turn and the phalanx of three returned to base amidst a flurry of expletives from their spearhead. Apparently, after a few metres of regulation, eyes-front march, Prabir Babu suddenly realized that he was alone, all alone in that wide, wide enveloping darkness. The about turn, very sensibly, was immediate.

When tempers cooled somewhat, I realized that something had to be done. I requested two men, Subrata and Prasad, to accompany me into the cover of bucolic bamboo and banana. By forest department standards, we were very well equipped indeed! Subrata was carrying a single-barrel departmental shotgun and a pocketful of No. 6 cartridges of uncertain vintage, adequate enough to frighten a covey of partridges but nothing larger. Prasad was carrying the most recent acquisition of the department, a .22 bore rifle with low velocity IOF cartridges, then in demand in shooting ranges for target practice. It had its utility, but it was certainly not the ideal piece of artillery to carry when expecting to meet a charging elephant in the dark: a three- or four-ton capsule of concentrated fury and mayhem.

When moving past a few village huts towards the clusters of bamboo and banana, I realized that the rather pleasant, soft moonlight made things worse. Things were made more ghostly by shadowy clumps of bamboo, patches of dense darkness at their base; on the other hand once we entered the canopy of trees and bamboo in a tunnel of darkness, visibility improved in the thin light of the hazy moon swimming placidly across the sky. It was the dark patches under the trees and clumps of bamboo impermeable by moonlight which now caused the trouble. Soon we realized that we were not alone in the wide expanse of grey and dark speckled with moonlight. This was a private orchard or plantation. Under almost every clump of bamboo were small groups of people. Of course they knew the place like the back of their hands, tree by tree, fruit by fruit. There were plenty of papaya trees around. Their fruits appeared to have a curious texture. With a beam of torchlight upon them the fruits revealed names of their owners, such as 'Babu', 'Rabi' and so forth, scratched upon their skin to discourage the sale of these autographed fruits in the local market after unauthorized collection. The absurdity of the situation was overwhelming: here I was in tweedy party-wear, following a large herd of elephants in the middle of the night in an orchard full of autographed papaya. My leather-soled shoes, however appropriate they were in the city, the undulating ground covered with loose soil vanquished them. In fact, the term 'undulating' is a misnomer here. The whole place was criss-crossed with deep and wide drains, presumably to prevent waterlogging. I kept slipping down one side of each trench we crossed as two sturdy sons of the soil pulled from behind to keep me from falling on my face, and after reaching the bottom of the pit they pulled me up the other side—certainly not a heroic spectacle. The light-and-shadow in the moonlight created strange effects. The dry drooping banana leaves looked like the sloping backs of elephants with wrinkled skin covered with dust. We were acutely aware all along that we could be ambushed from behind by a male lurking in one of the shadowy patches. When a herd is on the move such males usually trail behind in search of a stolen copulation, as the leading bull is usually with the herd lording it over his temporarily acquired harem. However, proceeding very slowly, trying to look, Janus-like, simultaneously both ahead and behind, we reached a small open space in the orchard. The villager leading our party lifted a hand and motioned us to stop. We froze at once. Immediately, we could hear the distinct throaty rumble of an elephant. We could not make out the animals separately, but we thought we could see some movement, a clump of bamboo slightly changing its place, a cluster of banana trees shifting ground. I asked Subrata and Prasad to lob two firecrackers, one immediately after another, keeping the .22 poised in my hands. There were some crashing sounds through the trees and bushes, a couple of squeaks and then silence. We came back to our vehicle and found the minister waiting for us. It was well past three in the morning.

We rushed back to the Delhi road and then went some way down the Durgapur Express Highway connector. The herd had not crossed either of the roads. There was now the daunting possibility that the herd had moved south towards Chandannagar and Srirampore along the western side of the highway. But we had to leave that for the next day.

At last, in the early hours of the morning, I rang my doorbell. More daunting than the elephants was the fusillade my wife unleashed. She had been informed around four in the afternoon over the phone from Writers' Building that I might be delayed returning home; the clerk had offered no further explanation. When I stayed away all of New Year's Eve and returned only in the morning in what appeared to her to be party dress, she assumed that I had been out tomcatting all over the city the whole night.

POSTSCRIPT

On the morning of 31 December, the herd, cajoled no doubt by the villagers from the west, entered Chinsurah town. There was no barrier of lorries then; in fact National Highway 2 had been left totally unguarded. As forest officials rushed with hoola parties to take charge, journalists had a field day photographing the herd passing through the narrow lanes of the town. The New Year opened with photographs of the herd and banner headlines hogging the front page, making journalistic history. Nobody however wrote of the night before as nobody had been around to record the night's events.

20

A Collar too Tight
East Medinipur, 1995

The forest department of West Bengal wanted to radio-collar an elephant in Medinipur in southwest Bengal as a part of a project that the Wildlife Institute of India was carrying out in that part of Bengal adjoining the Dalma Hills in Jharkhand. Its object was to survey and assess man–elephant conflict that had been raging in the eastern part of the region since 1987–8. It wass impossible for me to stay behind and the forest department let me come along.

We had with us two departmental koonkis, brought over from North Bengal and three on hire from Assam. The ground staff of the forest department reported the presence of a herd numbering about forty in the Nayabasat forest range of the East Medinipur forest division. Some of the koonkis were taken there by lorry; others stationed nearby were marched to the pre-arranged spot of assembly. A.K. Raha, then Conservator of Forests, Wildlife Circle, led the departmental team.

Quite a crowd had already gathered at the point of assembly by the time we reached. The place was all hustle and bustle with hundreds of onlookers milling around. The actual field team distributed itself on the koonkis: the radio-collaring team was on one; Subrata Pal Choudhuri and Gopal Tanti, the tranquillization experts of the forest department, were on another along with Dr V. Krishnamurthy, recently retired from the Tamil Nadu forest department. He would look after the veterinary side of the operation. I was on one koonki with two others; Mr Raha and his staff shared the remaining two koonkis. This time the team had gone quite high-tech: there were walkie-talkie sets with each koonki to help us keep in touch in dense cover. Following the koonkis there were about a couple of hundred of local heroes on foot who just refused to be denied the entertainment.

We entered the forest in a disciplined Indian file. Soon, in the dense cover, we gratefully noted our foot escort melting away. The wild elephants had already smelt the koonkis and were on the move. Helped by the mahouts from Assam, all expert trackers, we embarked on the herd's trail, hoping to catch up with them before long. The twigs on the ground were freshly broken: the leaves attached to them had not yet drooped; the grass was still flat in the semi-circular depressions in the ground made by the elephants' feet; the dung boluses on the ground were still moist and glistening despite being under the sun in some cases.

The tracks diverged and inevitably the team of koonkis split up. My koonki was driven by the most experienced phandi (elephant nooser) in the Assam team; our companion was another koonki from Assam. Our team of two koonkis was on one trail; the rest, including the radio-collar team and the darting party, took the other. Our excitement mounted as the trail already hot became scalding by the moment. Our mahout stopped abruptly and pointed his ankush to the ground. There was a puddle of urine on the ground, still frothing. We were really close. We remembered our walkie-talkie and activated the set to warn our companions that we were close to the herd— not a smart thing to do, really, when close to elephants with their exceptionally sharp hearing. The instrument crackled and a hoarse whisper in reply came from Mr Raha that they had actually come up with the herd. We were asked to switch off our set at once and stay absolutely quiet. They were trying to identify a suitable bull in the herd for tranquillization preparatory to radio-collaring.

We turned our koonkis half round to face a small clearing in the forest, and awaited developments. We had no idea where the other part of the team was. After a few minutes, leaves started rustling at a distance on our right, though not a leaf stirred where we were. The rustling grew louder and drew closer as if a cyclone was tearing down a narrow lane through the forest. We were, like Othello, perplexed to the extreme.

Before long the top of two or three elephants' heads appeared over the low sal coppice on our right. Soon some elephants emerged from the green screen on our right, followed by a lot more crowding behind them. They scampered across the scrubby opening in front of us, just like a herd of panic-stricken, stampeding cattle. On our two koonkis we were like the occupants of a box in a theatre. Before the herd disappeared in the forest to our left, we spotted two young tuskers among them, eminently suitable for radio-collaring.

While we were wondering if we should follow the herd or wait for the rest of the party who were obviously on the trail and had caused the stampede, there was another rustle and the sound of breaking twigs on our right. The massive head of a very large tusker appeared over the low line of the forest cover. An unhurried picture of regal dignity, a great tusker stepped out into the open ground, obviously on the trail of the herd. Moving about fifty metres behind, all its attention was focused on the retreating herd. About fifteen metres separated us from the great animal. That I was on a young tusker did not make me any happier. Sexual jealousy between two males, however disparate in age they might be, could alter situations completely. We went as still as carved relief on stone fervently hoping the great tusker would not notice us. What massive tusks! It was a gigantic animal, clearly at least a ten-footer, or even more. Our koonkis, around nine feet tall, looked positively puny compared with this bull. A sideways swipe of its massive tusks would have knocked out even the largest of our koonkis instantly.

Our koonkis stood as rock still as us, without a tremor in their body or flap of their ears, and, of course, without any vocalization. It was a demonstration of the temperament of ideally trained koonkis and their mahouts: not all koonkis and mahouts keep their cool when close to large unattached bulls.

Soon after we saw the other koonkis with their riders emerging on the trail of the herd. The party was as excited as we were. They too had spotted the two young bulls in the herd and were all for setting off immediately on the herd's trail. They had not seen the great bull and when we told them about the greater quarry just ahead of us, they politely refrained from expressing their disbelief.

We now proceeded on the trail together. Our friends still had eyes only for the young bulls they had seen, and seeing, as they say, is believing. Within a short time we came up with the rear of the herd. After scampering across the open, scrubby ground the herd had resumed its slow, normal pace upon regaining cover. One of the young bulls acting as rear guard was about twenty metres or so ahead of our two leading koonkis, one of them carrying the darting party. They fired their dart gun. The dart glanced off the leg of the animal. We could see clearly the drug in the syringe spraying out, and realized that it had been a miss. Once again the herd on the run.

Just then I saw tub-sized footprints branching off to the right. I was somehow able to persuade the party to abandon the herd for the time being and follow the tracks of what was clearly 'the grandfather of all tuskers'. Within a few minutes we came up against the big fellow who moved unhurriedly ahead of us, without taking recourse to any undignified scampering. In due course a dart was fired at its rump, which, unfortunately, hit it too low. Missing the fleshy part of the rump, it hit the right hind leg in the lower part of the thigh, where the vascular system is not dense enough for the drug to take quick effect. The tusker quickened its pace. We hurried after it with all our koonkis, but lost sight of it. Within a matter of fifty metres or so we were out of the patch of forests and on cultivated fields. Opposite us, beyond the paddy fields, was the embankment of one of the irrigation canals of the Kangsabati irrigation project. Hard, bare ground fringed the forest. We could spot no footprints; but all was not lost yet. There were shouts from a group of villagers bunched up on the far end of the embankment stretching to our right. They were gesticulating frantically, pointing to our left. We turned left. After negotiating a tongue of forest sticking out between the open ground and the embankment of the canal, we saw the tusker striding along the embankment, moving away from us. All attempts to close the gap between us and our quarry proved futile, though the mahouts were trying to goad their mounts almost to a run. One stride of the tusker appeared to equal two strides of our koonkis.

The drug had not taken effect yet, obviously. Our one fear was that when the drug did take effect, the tusker might roll down the embankment into the canal. There, paralysed and unable to move its trunk under the effect of the drug, it might get asphyxiated in the water of the canal.

We got on to the embankment and found to our relief that the canal's bed was dry. The tusker was well ahead of the koonkis of the darting party; about twenty metres further behind trailed our koonkis. Suddenly the tusker stopped, swaying. The drug—after nearly thirty minutes—was at last doing its work. We were praying the tusker would not fall into the canal but on the high and level field side of the canal, which would make radio-collaring of the animal just that much easier. Providence ignored our prayers; it continued swaying, and then rolled down to the canal bed to its right. In a few minutes we reached the spot where the tusker had toppled over to the bed of the canal, a good fifteen feet or so below. Lying there, its massive head resting against the inner lining of the canal, the tusker was a majestic and awesome sight.

Everyone dismounted and scrambled down the paved lining of the canal, now looking as if an avalanche had torn down it. We had not been able to take any photographs thus far because of our fast-moving, shaking mounts. My job now, I felt, was creating a photographic record of this great hero of the wild. Instead of getting off my mount I asked the mahout to take me down to the dry bed of the canal. From my elevated position on the koonki, I thought I would get good photographs of the whole process of putting a collar with a radio transmitter around the animal's neck. I began clicking away, sometimes urging the mahout to go closer, sometimes to step back or to move slightly to one side, just like a society photographer searching for the ideal angle when photographing royalty. Looking at the drugged tusker it did seem to me that we had committed lesè majesté, so overwhelmingly regal was the reclining figure of the monarch of the forest.

From the corner of my eye I noticed water entering the canal over the dyke on my right without quite registering its implications, dismissing it as an irrelevant distraction from the immediate and all-important photographic venture in hand. I was just asking the mahout to manoeuvre the koonki slightly to one side, my eye firmly fixed to the viewing mirror of the camera, when I noticed that people were moving away from the sleeping tusker. Gopal Tanti, who was also leaving with the rest, called out to me to leave at once. He said the tusker had just been given Revivon, the antidote to the tranquillizing drug, and was expected to recover consciousness and be on its feet in a few minutes.

There was no time to enquire why the programme of radio-collaring had been abandoned so abruptly. I confess my immediate concern was self-preservation. In a sudden reversal of roles, I felt like a trapped animal. The two inner sides of the canal, paved and steeply inclined, were the two long arms of a U, the wall-like dyke making up the base. We could never hope to run away along the open side of the U if the giant came rushing at us: we had had a demonstration of its speed. This was a trap, if ever there was one. The wild tusker was still half-inclined on the near long arm of the U. That my mount was also a male did not help matters. The last thing we wanted after the revival of the wild tusker was a display of male-vs-male aggression. I was never much of a hero, but rarely had I felt less like one.

The untried, paved and steep lining of the canal on the far side could be difficult for our mount. My mahout wanted to try that route, but I thought our mount just might refuse to go up that way, and then it would be too late to try another, given that the tusker was about to wake up. I decided that it would be safer to go up the same way that we had come down, though it was uncomfortably close to the still-recumbent tusker. Everyone else watched from the rim of the canal, shouting advice. We started climbing up close to the tusker's body, the mahout plying his ankush with all his panic-driven strength. No sooner had we reached the lip of the canal and the company of our friends, than the few minutes Gopal Tanti had given us were over. The wild tusker had heaved himself up and was standing in the water that was already ankle-deep and rising rapidly.

The revived giant took it easy, sprayed itself with water from the canal, then slowly sauntered to the far side of the canal and climbed up to safety and freedom. I was glad that I had not taken the mahout's suggestion to climb out of the canal from that side. The wild tusker had chosen exactly that route.

The dissection of events started. Why was the radio-collaring not done? The Wildlife Institute of India had calculated the required size of the collar on a formula devised by some American zoo. With our tusker it fell short by as much as a metre. The WII team had arrangements to extend the size of the collar, if necessary, by fixing an attachment. Unfortunately for us, the Kangsabati authorities had decided to release water into the canal just at that moment. This left us no time to attach the extension to the strap of the collar. The drugged animal, unable to move its trunk for air, might have drowned in the canal's water. There was no alternative to reviving the animal with the antidote immediately; hence the abandonment of the experiment midway.

POSTSCRIPT

The year after, broadly in the same area, a young tusker picked up a fight for dominance with our grand old patriarch. The young tusker won the fight and in the process inflicted a terrible wound on the older animal. Subrata rushed from Calcutta to sedate the animal chemically so that the veterinarians could treat the wounds. He had no difficulty in identifying our old friend. He climbed up a tree next to the sedated animal and with a weighted measuring tape found that the exact height of the animal was 10'4" at the shoulder. I think this was the first time that a standing wild elephant of this size was measured. Our old friend died, three days after his fatal combat. Natural selection had had its way.

21

Elephant in a Kraal
West Medinipur, 1995

The West Bengal forest directorate has attempted various methods to contain depredation by elephants in southwest Bengal. The first was chasing herds away with hoolas, flaming brands wielded by trained men. The practice was discontinued in West Bengal when it was found that hoola parties were actually causing more damage to crop. Herds, when chased, fan out when fleeing instead of following their normal practice of moving in single file, thereby increasing the area of damage, trampling swathes of crop. Hoola parties also have limited success with unattached males, which tend to be aggressive when a party with flaming brands closes in on them at night. Bhim Mahato, the leader of one such hoola party, was killed by an elephant in the course of his work. I do not know the exact circumstances of his death; but it is likely to have been caused by a too daring and too close an encounter with a lone adult bull.

Erecting energized fences all along the eastern border of Dalma to prevent the entry of elephants from Jharkhand into southwest Bengal was the second method. This was given up after two or three years, as it was found to be impractical. It has not been possible so far anywhere in India to maintain long stretches of energized fences, even over flat ground. In broken terrain, elephants can slip through under such fences. One problem is the pilferage of fencing wire. Villagers also interfere with the fence to let their cattle into the forest for daytime grazing, and then don't always restore the fence to its original operational position. The present perception is that such fences cannot be maintained without the cooperation and participation of the local people, say, through forest-protection committees, eco-development committees, panchayats and so forth. The current understanding is that a fence should be put up by the forest department only if the local bodies take the responsibility for its subsequent protection and maintenance.

In 1984, the state forest department put up a proposal to the Government of India to capture elephants in southwest Bengal: the intention was to reduce their number by half in a phased manner. The plan included the capture of adult animals by tranquillization followed by training in the kraal method* as practised in Tamil Nadu. The forest department had Dr V. Krishnamurthy, recently retired from the Tamil Nadu forest department, for training in the kraal method. West Bengal, Northeast and North India have no expertise in this. There would also be a team from Assam to capture and train elephants around seven feet high in the traditional Northeast India method. Capture and training in both the methods would go on concurrently.

The immediate problem, however, was with a small tusker, about eight feet in height, in the West Medinipur forest division near Jhargram. It had killed three people rather brutally, tearing its last victim limb from limb. The Assam team had not arrived yet, but three large departmental koonkis were already in Medinipur, brought in especially from North Bengal, to tackle adult animals after tranquillization. The departmental koonkis were stalled in the grounds of the Bandarbola forest beat office. The mahouts had prepared thick, hand-twisted jute ropes to tie up the large animals after their capture by chemical immobilization. Subrata Pal Choudhuri had arrived with

---

*A kraal is a chamber made of round logs, approximately ten feet square; in the kraal method, a freshly captured animal is put inside the kraal and trained without ropes, thus avoiding rope galls, which can be severe in large animals.

dart guns and chemicals. Subimal Roy, then Chief Wildlife Warden of the state, was to supervise the operation. So, all the main actors were assembled there, the *mise en scène* in place, waiting for the action to begin. I was fortunate enough to be present, as one of those who are around and help to swell the progress, as an 'extra'.

Information came from the Divisional Forest Office, West Medinipur, that the rogue tusker had chased a man the previous day, a few kilometres south of Bandarbola, where, by the happiest of coincidences, the koonkis were.

The day after, when we gathered at Bandarbola, we were immediately greeted with the news that the animal had been spotted the previous night at a place south of the highway between Kolkata and Mumbai (National Highway 6), a good 30 kilometres or more from where we were. Three hired lorries arrived to take the koonkis to the spot. Koonkis are temperamental creatures, and one has to know one's koonkis on such occasions. Some, such as Jatra Prasad, would enter lorries head first; some, such as Chandrachud, preferred to back in to a lorry, which made unloading that much easier as the animal could see and feel what it was stepping down to; some on the other hand just refuse to get into a lorry at all. In all cases, whether loading or unloading, the height of the lorry's floor from the ground has to be right. Ramps sometimes have to be built or improvised on the spot for the purpose.

We reached the spot on the highway, somewhere between Balibhasa and the Jhargram road crossing, in an hour's time. The unloading of the *koonkis* was smooth. The ever-watchful villagers and local forest staff informed us that the rogue was at that time another kilometre and a half further south. We proceeded to the place on elephant-back. Chandrachud was leading, carrying the darting party consisting of Dr Krishnamurthy and Subrata Pal Choudhuri. Mr Roy along with the Divisional Forest Officer and his staff was on Rajkumari, and I followed behind on Jatra Prasad along with some forest staff. A big crowd had assembled on the narrow road. They directed us to the exact spot on the right of the forest track. The rogue, they informed us, was now with a family group, which we later saw numbered four: the matriarch, calves of different sizes, obviously of successive litters, and our quarry, the chakna, the tallest and the oldest among the siblings, though slightly shorter than the matriarch.

The chakna was launched on its roguish career when a huge tusker joined the herd one day, picked a fight with the chakna and drove it away. Fortunately for us, the big tusker, usually not aggressive to man, was not with the group then. After a fortnight's dalliance with the leading female of the group it had moved away, presumably in search of new pastures. With that threat removed, the chakna had rejoined the group.

This, for me, threw a new light on the herd behaviour of elephants. It has long been accepted that pubescent male calves are expelled from the herd as nature's way of preventing genetically undesirable in-breeding. It has also been assumed that the family group, perhaps the matriarch herself, expels the young male. Here was an instance of how things actually happened—an adult male seeking to mate sees in the young male a potential rival and expels it from the proximity of the female of its choice to ensure mating success for itself. This also explains why an expelled animal frequently turns aggressive.

After expulsion from the herd the young elephant's gregarious instincts seek the company of other males where the pecking order is quickly established and accepted. The process of learning by emulation then can continue in the association of an older animal. A long process of learning is thought to be natural to a species with a long lifespan. Thus old animals frequently have young males in tow; these pairs, called *askaris* in East Africa, are not be confused with a 'consenting adults'-like situation among humans. This behaviour pattern among elephants could explain the formation of male groups or bachelor parties.

The initial aggression in the elephant immediately after expulsion from the family group or herd becomes unacceptable to humans when the animal starts venting its aggression on them rather than other elephants. The animal then has to be removed from the wild by capture and denied the freedom to move around at will. Earlier the practice was to shoot such an animal down. The practice still continues, though this teenage aggression is often a passing phase in the animal's life.

As we proceeded carefully towards the elephant group, we could hear a loud buzz of conversation coming from the road left behind. Animals in Medinipur are so blasé about human proximity that the group seemed unperturbed by all the voices. From a distance of about twenty metres Subrata fired his first dart frontally. It must have hit the animal somewhat high in the cranium, missing the major vascular system of the trunk. Even then, despite the sharp 'ping' of the .22 cartridge propelling the dart, there was no panic in the group. Medinipur elephants take such noise in their stride, even the blaring horns of big transport lorries, so close are they to humans all the time. Chandrachud closed in, but the group was indifferent. The chakna by now was showing some mild reaction to the drug. It was resting its trunk on the back of a sub-adult female in front, suggesting that its trunk was getting limp. Subrata shot off the second dart. The group still showed no sign of agitation or any inclination to move away. After a few minutes the darted animal sank down on its legs and rolled over to its side. The group was still close by and had to be shooed away with the bursting of crackers and shouts before the team could begin work on the unconscious animal.

A little later Jatra Prasad and Chandrachud flanked the prone chakna, as ropes were tied round the neck of the drugged animal on the ground, and fixed to the girth belts of the two koonkis. They were ready to haul the captive away to the waiting lorry. The animal was revived with the antidote and was literally fighting fit in a few minutes. The leads from the neck of the captive to the girth belt of the koonkis were very long, following the practice in South India. The long ropes gave too much leeway to the captive animal when being pulled along a narrow, forest-lined path. Tied closer to the two koonkis, as is customary in the Northeast, the control on the captive animal is much tighter. We fell back on the Northeast Indian system in the final stage of hauling the animal as the chakna put up a big fight, becoming almost unmanageable.

In the beginning, however, taking advantage of the slack, the captive animal at one point almost escaped into the trees on the left of the track. Tugging the rope, in order to enter the forest, the animal turned round to challenge Chandrachud. Chandrachud under its mahout's command, faced the chakna head on in a position called *chowdanti* (*chow* = four; *danti* = tusks) in Lower Assam. It did not take Chandrachud more than a minute or two to teach the chakna who the boss was: after all Chandrachud was more than a foot taller than the chakna. It is amazing how quickly the height and the bulk of an animal establish its rank in the pecking order among elephants. From that moment the chakna became a docile chela of Chandrachud who, in turn, became protective towards it.

With the chakna continuing to be so violent we shouted to Subrata to give it a topping dose of Rompun, made of Xylazine, a muscle relaxant. We found out late, much too embarrassingly late, that in all that excitement, there had been an awful mix up. Wise from past experience, Subrata was carrying two extra syringes for use in an emergency: one loaded with Rompun and the other with Immobilon (a mixture of Etorphine and Acepromazine). In the confusion, the chakna got a second syringe of Immobilon.

After a few hundred metres we came to the forecourt of a village school. The captive passed out. It was slightly past midday and very hot. The ambient temperature was really not suitable for using Etorphine, which raises the body temperature of an animal receiving the drug. We doused the recumbent body of the drugged animal with buckets of water brought in relay from the school's tubewell. The elephant showed no sign of animation. Assuming it was a case of an overdose of Rompun, we decided to wait it out, not having with us Yohambin, the specific antidote for Rompun. We were so sure of this that I slipped out to a dhaba and had a chai followed by a snooze on a string charpoi alive with bedbugs. Recovering somewhat after an hour or so from my state of exhausted stupor—I was recovering from a recent bout of typhoid then—I hitched a lift on a departmental vehicle carrying back lunch packets for the field party. Reaching the school ground, I was worried to see the captive still on the ground, Jatra Prasad and Chandrachud standing on either side, its neck ropes still tied to their girth belts. Dr Krishnamurthy, confused about the problem, kept checking the supine animal's blood pressure and body temperature, and examining the fixed, open eyes of the animal for involuntary reflexes to light. A cardiac arrest seemed imminent. Jeeps rushed to the nearest town for supportive medicines and equipment. Intravenous glucose and a saline drip were started.

In course of time it became dark. There was, of course, no question of hauling the animal over 35 kilometres at night to the kraal near Bandarbola, even if the animal revived. Mr Roy and I rushed to the authorities of a mission school across the highway for permission to keep the koonkis and the captive animal in the school grounds for the night. Generously, the school authorities agreed to have a proven killer weighing nearly three tons on their premises as an overnight guest. Pressure lamps created pools of bright light on the dark school ground against a black backdrop. Coming back to our old school compound we found the partly lit, ghostly shapes of Jatra Prasad and Chandrachud, munching peacefully on plantain stalks specially brought for them, floating immaterially. There was a sudden spasm of animation among the darting team as they discovered the problem was a double dose of Immobilon. Now a dose of the antidote was pushed in intramuscularly. Everybody was on the alert, including Jatra Prasad and Chandrachud, and their mahouts. We moved a few paces back to widen the empty space around the animal lying on the ground, expecting aggression after revival, as we had experienced earlier.

Within a few minutes of receiving the antidote for Immobilon, the free ear of the animal started flapping, each flap sending waves of relief and joy through us. Another five minutes, and the animal was on its feet, still in a daze, without any sign of the aggression or belligerence that we expected. We started to march it off at once, tied, tightly now, between two koonkis, to the pre-selected site in the grounds of the mission school.

It would have been very difficult to march three elephants abreast, the captive flanked by the two koonkis, along the narrow road by the canal joining Bandarbola with the highway. The main road connecting the highway with Jhargram was ideal for our purpose, but hauling a killer elephant along that road on a normal day with its heavy traffic would have been very hazardous indeed!

Fortunately for us Mamata Bannerjee unwittingly came to our rescue: it was Independence Day and she had called one of her unpredictable bandhs. The road was clear of all vehicular traffic, and even the most rabid supporters of the bandh did not care to inform a killer wild tusker that the bandh did not allow movement on roads.

My fever-ravaged body was trying to recover from the exhaustion of the previous day. As a result I overslept in the morning. Mr Roy and Subrata had not had the heart to wake me up. The party had already left when I woke up and when I reached the junction of the Jhargram road and the Bombay road, I saw that things were already on the move. People had gathered at the crossing for a spectacle that Medinipur, or indeed the whole of southwest Bengal, had never seen before. Soon I could see the bobbing heads of the koonkis and the captive tied between them. Now and then the outline of Rajkumari could be seen behind them.

It was a triumphant and jubilant parade. First came almost the entire forest staff of the division, nobody willing to miss that moment of glory. They had lived with the problem for weeks; now that it had been overcome, it was certainly a moment of triumph. They were in a bunch in front, enthusiastically clearing all the obstacles on the way, mostly men on bicycles who had come from places as far away as Jhargram town to see the fun. The staff seemed to enjoy every step of the long kilometres they walked on that excruciatingly hot day. Next came the captive between Jatra Prasad and Chandrachud, with Rajkumari acting as the rearguard. After a gap of twenty-five metres or so, followed departmental vehicles carrying the supervisory staff. The thirty-five-kilometre-long route march went off without a hitch because of the meticulous planning by Subimal Roy. Every ten kilometres of the route, resting places had been created. Some cut fodder was waiting there for the elephants, as well as water for spraying the captive and the koonkis as protection against the high ambient temperature. Reaching Bandarbola, we turned right and followed a narrow village road to the grounds of a forest Beat Office where a kraal, specially constructed for the occasion, was waiting to receive its guest. After another half an hour, the captive was in its temporary home.

POSTSCRIPT

The captured animal was amenable to training, and put to service, but died after two years from an infectious disease of the feet which had turned septic. There were other casualties that season among the elephants captured. One captured tusker, after being transferred to North Bengal, escaped to the forest. Two died after being put into the kraal. The capture of a young female just seven feet tall was a success. She was captured and trained in the traditional methods of northeast India and patrolled the forests, carrying tourists at the Garumara wildlife sanctuary in North Bengal.

I learnt a few lessons during this operation.

1. The only acceptable way of capturing and training adult animals (seven feet and above in height) is tranquillization followed by training in kraal.

2. The Karnataka practice of allowing the freshly captured animal to cool down first in a kraal, and then training it outside the kraal tied between koonkis in the Northeast India manner, a process that takes about three months, is preferable to training in the kraal itself, as practised in Tamil Nadu.

3. A killer elephant can be domesticated and trained, and need not necessarily be eliminated.

4. The Northeast India practice of capture and post-capture training is preferable for sub-adult animals (seven feet and below), as this reduces the period of stress. It takes about fifteen days in the Northeast India method to complete the first phase of the training: in which it learns four basic command words and gets used to carrying its own fodder. In the South Indian practice, it takes about three months to complete what is considered the first phase of training.

5. Muscle relaxants like Rompun can be used in the initial few days of enkraaling when the animal is at its most violent.

22

The Indian Elephant
A Window on the Future

While the Mughals caused the number of elephants in the wild to decrease by excessive capture, the British encouraged bounty-hunting on a large scale. Both the capture and destruction of elephants went on unchecked in British India up to 1879. Sport hunting was an added hazard.

Oddly enough, capture, which decimated the number of elephants in the wild, also strengthened their long-term chances of survival, and encouraged bonding between man and elephant. More than anything else, it is this bonding that has ensured the survival of the species in a country like India with enormous anthropogenic pressures on elephant habitat. A comparison with Africa, where there has been practically no capture and domestication of elephants—and therefore no human bonding with elephants—for more than two thousand years, is instructive. As a consequence, the culling of elephants in Africa by the thousands annually has not led to any popular resentment.

After 1972, the pendulum in India started swinging the other way. The elephant was elevated to Schedule-I of the Wildlife (Preservation) Act in 1977 and became a protected species. India's Project Elephant was launched by the Government of India in 1992 after a detailed discussion of the strategy to be adopted to ensure the long-term survival of viable, large elephant populations in India. In the 1980s, with the ruling dictum that 'small is beautiful', a new programme for preserving biodiversity gained currency. The idea was fine on paper: varieties of gene-pools were to be preserved in small puddles. In reality, these little gardens of Eden could not cope with burgeoning anthropogenic pressures and the compulsions of a stagnant economy. The urge to survive got the better of the urge to conserve.

It had been thought that religious groves would have a special role in this context. People would of their own volition preserve these plots on religious grounds. These groves were indeed important as they were often pools of relict and endemic species of vegetation. But again the urge to survive took precedence. The religious groves in the Khasi Hills are a case in point. We have a list of them in the early gazetteers; but few of them have survived. The Balphakram plateau in the Garo Hills, apart from its large area, has the added advantage of being a significant place, spiritually, for the Garos. This is where the souls of good Garos go after death. This would seem the strongest possible religious taboo against interference with nature on that plateau. Yet the endangered *Aquilaria malaccensis* (agar) trees were systematically worked on its slopes with the encouragement of the authorities, local as well as district level. The wealth of endemic bamboo in Balphakram, the principal building material used in the Garo Hills, has not escaped the axe either.

All our sanctuaries were set up on small areas selected for the protection of species with limited home ranges. Some contiguous protected area networks provide protection over very large tracts: such as the Mudumalai–Bandipur–Nagarahole complex in peninsular India; the West Bengal and Bangladesh Sundarbans taken together; the chain of protected areas extending from the Manas tiger reserve to the adjoining Buxa tiger reserve and the Jaldapara wildlife sanctuary in West Bengal. But these were accidents of political geography, and large sanctuaries are by no means the norm in India.

Project Elephant realized in its preparatory stage that sanctuaries set up on 'core-and-buffer' principles would not work for a long-ranging species like the elephant. Not a single sanctuary in India, by itself, protected the entire home range of an elephant population. Wild elephants range over such a large area that in the Indian context it would be impossible to exclude all human activity from it. Thus the management of these ranges cannot be exclusionary in approach. Project Elephant has, therefore, adopted protection and management of the entire range as the guiding principle of its conservation efforts, allowing such controlled exploitation of the habitat by man as may be compatible with the demands of elephants. As a corollary, one of the prime tasks of the Project is to reduce man–elephant conflict; for it realizes that elephants cannot survive in the face of man's antagonism.

Eleven such ranges were initially identified in the country: five in Northeast India; one in North India; one in Eastern or Central India; and four in South India. For administrative convenience the Project initiated a programme in 2001 to notify the state-wise components of each range as separate elephant reserves, which by December 2003 numbered 26 and spread over 61,000 sq. km. of habitat with about 20,000 elephants. Some more reserves are being planned and some links are still missing in the chain of identified elephant ranges which need to be brought under intensive management as elephant reserves. Elephant reserves, it should be remembered, are not sanctuaries or national parks; they are management units set up by executive order just as tiger reserves are, and do not enjoy a separate legal status of their own.

Ten years after the launching of Project Elephant, I must admit the Project manifests the typical 'Hindu rate of progress', as economists once used to call the rate of India's economic growth. Even basic ten-year management plans for these reserves are still to emerge. But one does not want to strike too pessimistic a note. The geographical range of the species is far greater in India than in most Asian countries, though some excellent work has recently been carried out in Borneo, and in Sri Lanka in the past.

Mitigation of man–elephant conflict has been recognized as one of the main tasks before the Project, which attempts to evolve a *modus vivendi* for man and elephant to coexist. It is also recognized that the bond that has developed in India between man and elephant over millennia, and not mere administrative resolve, is our greatest strength in our effort to conserve the species. The foundation of this bonding rests on man and elephant living together for thousands of years. Thus domesticated elephants, their welfare and proper management, feature prominently in the project's agenda, which includes training of mahouts and proper veterinary care for elephants.

All said and done, the question remains: how many elephants can India afford? As the human population increases, so does the elephant population. Do we look for the *maximum* or *optimum* number?

On paper, there is a positive trend in the enumeration of elephants. The figures in 2002 were 27,413, an increase from about 18,000 in 1993, suggesting an upward movement. However, at the same time, there are definite indications that this may be a case of general decline combined with local overabundance. For example, elephants in the Barak Valley in Assam are nearly on the verge of extinction whereas elephant populations in North Bengal and Dalma in Jharkhand have shown an upward swing. We can add to this the number of domesticated elephants in the country, which was estimated to be about 3000 in 2003. A programme launched by Project Elephant is under way, of micro-chipping these elephants to identify permanently each one of them. Till 2003, about a thousand elephants had been marked with chips.

However, these figures can be misleading. No reliable census method is as yet available to us for the hilly and mountainous terrain where the bulk of our elephant population lives. Project Elephant is trying to devise a suitable census method with the help of the Geographical Positioning System. However, no uniform method for counting elephants is followed throughout the country. The 'total count' method has been practised in more than half the area of elephant country; sample count in quadrats in some, line transect counts in others, and dung count or water-hole counts in the rest. No reliable all-India database can be prepared on the basis of counts using widely different methods. And in some cases there are obvious anomalies: the case of Meghalaya, for example, where a drastic reduction has been recorded without any discernible cause.

Then there is the flip side of the question of increasing numbers of elephants: man-elephant conflict. Between 1991–2 and 2002–3, a total of 719 animals were poached, that is, an average of 60 per year. From 1998 to 2003, 72 elephants were killed and 7 were injured in railway accidents, out of which 41 were cows and calves. Most regrettably, this did not seem to worry or agitate the railways too much.

During the period 1991 to 2003–4, 2856 cases of human deaths were recorded. Enumeration figures of wild elephant population showed a steady increase from 1980 to 2002. During the same period, there were 721 cases of elephant poaching in India. All these cases of poaching, that is, illegal killing, were not necessarily for ivory; sometimes they were retaliatory killings of marauding elephants, caused largely by the inaction and apathy of the forest department. Between 1998 and 2004, 40 elephants died from poisoning and 186 elephants from deliberate electrocution. Neither poisoning nor electrocution is a respecter of age or sex of animals.

It is therefore necessary to have a clear idea of optimal density when at present the mindset of some conservationists appears to be leaning towards 'maximum density', which seems unsustainable in the present situation.

The situation in India perhaps has a parallel in Africa: general decline with local overpopulation. In 1993, at a seminar organized by the BNHS at Mudumalai, conservationists from Zimbabwe reported that they had evolved an equation/formula to determine the optimal density of animal populations in their forests, with the damage to tree cover as the parameter. The object was to quantify the culling necessary. It is, I feel, necessary to have in the Indian context at least two sets of parameters for determining optimal elephant density:

(a) Damage to forest cover and negative impact on natural regeneration.

(b) Man–elephant conflict, calculating the 'edge effect' or 'cutting edge' of forests, determined by how much human habitat is exposed to elephants, as well as elephant numbers and the tendency of certain populations to extend their home range, thereby increasing the 'edge effect' or 'cutting edge'. The more elephants increase their home range, the more their habitat coincides with those of humans, leading to greater possibilities of conflict.

When we look out of the window on the future, the view may not be gloriously rosy, but it is not one of unrelieved gloom either. There is every reason for lingering clouds to gradually drift away, towards a future that has room for both man and elephant.

Glossary

Ādāng (Garo)	:	Area of dry cultivation by jhuming (q.v.).
Agaru/Agar	:	*Aquilaria malaccensis.* Agaru (*agar ruh*) is the essential oil extracted from the agar tree.
Alpinia	:	*Alpinia allughas.*
Aman paddy	:	Autumn paddy.
Bhabar	:	Typical of the sub-Himalayan tract, a stretch of waterless, wide, dry porous soil, separating the foothills from wet terai land formation farther south.
Chharā	:	Small stream.
Chāknā	:	In Northeast India and among people dealing in live elephants in North India, tuskers are classified according to the shape and size of their tusks. Ideally, a chakna should have very short and very thick tusks. In common usage, however, it has come to mean an elephant with small tusks of no particular distinction.
Chārjamā	:	A square box put on riding elephants with wickerwork sides, supposedly not a proper howdah, either for ceremonial or hunting purposes. Traditionally, I am told, it was used to carry ladies of the Mughal court and had screens on the sides

and a canopy on top. This may not be right as many Mughal miniatures show the Badshah hunting dangerous game from such a box. It is possible that the English devised the standard nineteenth-century hunting howdah, where the shikari stood up straight holding to the front rail, for their convenience. Sitting cross-legged in the howdah was not their style.

Chiring (Garo)	:	Stream
Dāo	:	Short-handled chopper, an instrument for all purposes, without which a Garo would feel underdressed.
Dulshi	:	A bunch of thin ropes slung round an elephant's neck, acting as a stirrup for the mahout.
Gaddy	:	A part of an elephant's riding gear, a thick pad which is placed on an elephant's back and secured with girth ropes.
Gajbāg (North India)	:	Elephant goad or tamer. As Abu'l Fazl annotates the word, Akbar the Great coined it to substitute the original Sanskrit word *ankush*, which Akbar's Turkish courtiers tended to mispronounce as *angoj*.
Hoolā	:	Flaming brand
Jhum/Jum	:	Slash-and-burn cultivation, usually starting at the end of the dry season in March and the beginning of the wet in April.
Khair	:	*Acacia catechu*. Khair, *sissoo* (q.v.) and *simul* (Bombax) are the pioneering tree species in the sandy soil of riverine forests.
Khedā	:	A method of capturing elephants where a herd is driven (*kheda*) into a stockade.
Khupi	:	An elephant command word meaning 'stand firm with legs spread to withstand any pull, such as by a just-captured wild elephant.'
Kumeriah bāndh	:	Description of the body structure of an elephant. *Koomeriah* or 'princely', *bandh* or 'body structure' is the one most highly prized.
Koonki	:	Domesticated elephant trained to tackle wild elephants.
Machān	:	A platform in an elevated position.

Mahal	:	Demarcated area leased out for operation.
Māknā	:	Tuskless male elephants. In the Asian species, the females do not have tusks, nor do all males.
Māljuria	:	A term in North East India for a group of only male elephants, also called 'bachelor parties' in Africa.
Mālkhāna	:	Strong room, usually guarded.
Mār boat	:	Flat-bottomed, wide country boats, sometimes with power, used to ferry people, vehicles and heavy cargo across.
Melā shikār	:	Capture of sub-adult animals and calves by noosing.
Mongreng(Garo)	:	Short-handled Garo dao.
Muli bamboo	:	*Melocanna bambusoides.*
Musth	:	A condition in adult Asian male elephants when the temporal glands swell and exude a viscous, smelly secretion; the animal may become ungovernably aggressive. Chemical analyses have shown that the secretion is almost pure testesterone. This is the general rule, but there are exceptions. Female Asian elephants, usually when pregnant, may exude a small amount of such secretion for a very short period, but do not, as a rule, turn dangerous to man.
Pāik	:	Footguard.
Phāndi	:	Elephant nooser.
Pāttāwāllāh	:	The mahout's assistant whose special charge is collecting fodder for the animal. *Patta* = leaves and green fodder. They are also variously called *kabadi, mate,* or *meti.*
Pilkhānā	:	Elephant stable.
Sardāri koonkis	:	Leader koonkis, a term used in northeastern India: usually domesticated males of outstanding size and bulk, which play a leading role when bringing out freshly-captured wild elephants from a stockade.
Sissoo	:	*Dalbergia sissoo.* See *Khair* above.